M. P. CHERKASOVA

COLLECTED PROBLEMS IN NUMERICAL METHODS

M. P. CHERKASOVA

COLLECTED PROBLEMS IN NUMERICAL METHODS

Translated and edited from the Russian by
Dr. G. L. THOMAS
University of London King's College
and
Dr. R. S. ANDERSSEN
Computer Center,
The Australian National University

AKADEMIE-VERLAG · BERLIN 1972

М. П. Черкасова

Сборник задач по численным методам

Hochschulverlag, Minsk

Erschienen im Akademie-Verlag GmbH, 108 Berlin, Leipziger Straße 3–4

Lizenznummer: 202 · 100/562/72
Einbandgestaltung: Karl Salzbrunn
Gesamtherstellung: IV/2/14 VEB Druckerei »Gottfried Wilhelm Leibniz«,
445 Gräfenhainichen · 3795
Bestellnummer: 761 694 5 (5993) · ES 19 B 4
Printed in German Democratic Republic

CONTENTS

EDITOR'S NOTE

In order to make this translation as useful as possible, additional English references have been liberally inserted throughout the book. In addition, certain changes of structure and explanation have been made to the Russian text to improve its clarity. In many of the problems in the book, the notation $a(h)b$ is used to denote the values of a certain parameter. For example, see Problem 43 in Chapter 1. Its interpretation is: the set of values a, $a + h$, $\ldots$, $a + n\,h\,(= b)$ where $n = (b - a)/h$. For some of the problems of Chapter 2, the matrix A must be constructed as $A = C + kD$. For example, see Problem 38 in Chapter 2.

R. S. Anderssen,
Canberra, 1970

Chapter 1

THE APPROXIMATE SOLUTION OF NON-LINEAR ALGEBRAIC AND TRANSCENDENTAL EQUATIONS

Consider the equation

$$f(x) = 0, \tag{a}$$

where the algebraic or transcendental function $f(x)$ is defined and continuous in a finite or infinite interval.

A transcendental equation may have no roots, a finite, or an infinite number of (real and/or complex) roots, while an algebraic equation of degree m has m (real and/or complex) roots.

Often, the roots of (a) can be determined only after they have been localized, i.e., for any one of the roots, a sufficiently small neighbourhood containing only that root is specified. The position of this localized root must then be determined with a certain degree of accuracy.

If the function $f(x)$ changes sign in one of the intervals

$$[x^* - A,\ x^*],\ [x^*,\ x^* + A],$$

then x^* defines an approximation to the root of $f(x) = 0$ with accuracy A. To estimate the error in x^*, we can use the expression [1]

$$|x^* - \alpha| \leqq \frac{|f(x^*)|}{M},$$

where α is the exact root of (a), α and x^* lie in the same interval $[a, b]$ and $|f'(x)| \geqq M > 0$ for $a \leqq x \leqq b$. If $f(x)$ is a complicated function it may be difficult to estimate $|f'(x)|$. Hence, if the interval $[a, b]$ containing x^* and α is sufficiently small,

$$|x^* - \alpha| \leqq b - a,$$

can be used as a measure of the error. In particular, the successive approximations x_n obtained using Dandelin's method and the direct method of iteration (with $\varphi' < 0$), which are discussed in

§ 1 below, lie alternately on opposite sides of α and form a sequence of nested intervals. In this case two successive approximations to the root, i.e.,

$$|x_n - \alpha| \leqq |x_n - x_{n-1}|,$$

can be used to determine whether x_n has been calculated with the desired accuracy.

The following expression can be used to estimate the error in the approximation ξ to a complex root z^* [1]:

$$|z^* - \xi| \leqq \frac{|f(z^*)|}{M},$$

where $|f'(z)| \geqq M > 0$ for $z \in U(z^*, \xi \in U)$, and U is a convex set of points in the complex plane.

Below, we define various methods for solving $f(x) = 0$. We shall assume that the required root is $x = \alpha$.

For further reading, see Fröberg [2; chapter 2], Ralston [3; chapter 8], Moursund and Duris [4; chapters 1, 3], James, Smith and Wolford [5; chapter 3], and Berezin and Zhidkov [6; chapter 7].

1. Iterative methods for algebraic and transcendental equations of a single variable

THEORY

1. Newton's method {see [2; § 2.2]}

$$x_n = x_{n-1} - \frac{f(x_{n-1})}{f'(x_{n-1})} \qquad (n = 1, 2, \ldots).$$

The initial approximation x_0 must lie in a suitably small neighbourhood of $x = \alpha$ with $f'(x) \neq 0$ in this neighbourhood. This method can be used to find not only real but also complex roots. In the latter case x_0 must be complex.

2. The method of false position {see [3; § 8.3]}

$$x_n = \frac{x_0 f(x_{n-1}) - x_{n-1} f(x_0)}{f(x_{n-1}) - f(x_0)} \qquad (n = 2, 3, \ldots).$$

The initial approximation x_0 is any point in a suitably small neighbourhood of $x = \alpha$ such that $f(x_0) \cdot f''(x_0) > 0$. The second approximation x_1 is any point in this neighbourhood such that $f(x_1) f(x_0) < 0$ {i.e., $f(x_1)$ and $f(x_0)$ have opposite signs}.

3. Dandelin's method

In this method the above two methods are combined.

$$\left.\begin{aligned} x_{2n} &= x_{2n-2} - \frac{f(x_{2n-2})}{f'(x_{2n-2})}, \\ x_{2n+1} &= \frac{x_{2n-2} f(x_{2n-1}) - x_{2n-1} f(x_{2n-2})}{f(x_{2n-1}) - f(x_{2n-2})} \end{aligned}\right\} \quad (n = 1, 2, \ldots).$$

The values x_0 and x_1 are determined from the equations

$$x_0 = a - \frac{f(a)}{f'(a)}, \qquad x_1 = \frac{a f(b) - b f(a)}{f(b) - f(a)},$$

where a and b correspond to x_0 and x_1 of 2, respectively.

4. The direct method of iteration {see [2; § 2.5]}

$$x_{n+1} = \varphi(x_n) \qquad (n = 0, 1, 2, \ldots).$$

The function $\varphi(x)$ is obtained by transforming $f(x) = 0$ into the form $x = \varphi(x)$. For the iteration to converge it is necessary that $|\varphi'(x)| < 1$ in some neighbourhood of $x = \alpha$. The initial approximation x_0 must lie in this neighbourhood.

5. Bairstow's (or Hitchcock's) method for quadratic factors {see [2; § 2.3]}

A polynomial with real coefficients

$$f(x) = x^n + a_1 x^{n-1} + \cdots + a_{n-1} x + a_n$$

can be factorized as the product of polynomials of the second degree.

Hence, the problem of determining the roots of

$$f(x) = x^n + a_1 x^{n-1} + \cdots + a_{n-1} x + a_n = 0$$

can be reduced to that of solving equations of the first and/or second degree.

Bairstow's method for evaluating a quadratic factor

$$g(x) = x^2 + p x + q$$

in $f(x)$ is:

(i) Knowing the k-th approximations $p^{(k)}$ and $q^{(k)}$ to the coefficients p and q of $g(x)$ $(k = 0, 1, 2, \ldots)$ and putting $p \approx p^{(k)}$ and $q \approx q^{(k)}$, we divide $f(x)$ by $g(x)$ and $[g(x)]^2$ to obtain:

$$f(x) \equiv (x^2 + p\,x + q)\,L(x) + xP(p, q) + Q(p, q),$$
$$L(x) \equiv (x^2 + p\,x + q)\,L_1(x) + xR(p, q) + S(p, q).$$

(ii) Knowing $P(p, q)$, $Q(p, q)$, $R(p, q)$ and $S(p, q)$, we determine
$P_p'(p, q) = pR(p, q) - S(p, q)$, $P_q'(p, q) = -R(p, q)$,
$Q_p'(p, q) = qR(p, q)$, $Q_q'(p, q) = -S(p, q)$.

(iii) The solution of

$$\begin{cases} P_p'(p, q)\,\Delta p^{(k)} + P_q'(p, q)\,\Delta q^{(k)} = -P(p, q), \\ Q_p'(p, q)\,\Delta p^{(k)} + Q_q'(p, q)\,\Delta q^{(k)} = -Q(p, q), \end{cases}$$

determines the corrections $\Delta p^{(k)}$ and $\Delta q^{(k)}$, and hence, the next approximation:

$$p^{(k+1)} = p^{(k)} + \Delta p^{(k)}, \quad q^{(k+1)} = q^{(k)} + \Delta q^{(k)}.$$

Thus, the coefficients of the required quadratic factor can be evaluated to any given accuracy if the initial approximations $p^{(0)}$ and $q^{(0)}$ define a sufficiently accurate first approximation.
By extracting this quadratic factor from $f(x)$ using synthetic division (Horner's scheme) {see [2; § 2.1]}, the process can be continued inductively until all are determined.

6. Graeffe root squaring (or Lobachevskiy's) method {see [3; § 8.10]}

The advantages of the Graeffe root squaring method are: all the roots are determined simultaneously, and no preliminary estimates are required. However, the method only applies to polynomial equations:

$$P_n(x) = P_n^{(0)}(x) = x^n + a_1\,x^{n-1} + \cdots + a_{n-1}\,x + a_n = 0.$$

Denoting the roots of $P_n(x) = 0$ by $\alpha_1, \alpha_2, \ldots, \alpha_n$, it follows that

$$P_n^{(1)}(x) = (-1)^n\,P_n^{(0)}(x)\,P_n^{(0)}(-x)$$
$$= (z^2 - \alpha_1^2) \cdots (z^2 - \alpha_n^2).$$

Inductively, we define

$$P_n^{(r+1)}(x) = (-1)^n\,P_n^{(r)}(x)\,P_n^{(r)}(-x) \quad (r = 2, 3, \ldots) \tag{1}$$

and denote the coefficients of $P_n^{(r)}(x)$ by $a_j^{(r)}$ $(j = 1, 2, \ldots, n)$.

It therefore follows from (1) that for $j = 1, 2, \ldots, n$,

$$a_j^{(r)} = \begin{cases} (-1)^j \left\{ [a_j^{(r-1)}]^2 + 2 \sum_{k=1}^{\min(n-j,j)} (-1)^k a_{j-k}^{(r-1)} a_{j+k}^{(r-1)} \right\}, & (2) \\ (-1)^j S_j(\alpha_1^{2^r}, \alpha_2^{2^r}, \ldots, \alpha_n^{2^r}), & (3) \end{cases}$$

where $\min(n-j, j)$ = minimum of the two possibilities $n-j$ and j,

$$S_j(x_1, \ldots, x_n) = \sum_1^n{}_c \, x_{k_1} x_{k_2} \cdots x_{k_j},$$

where $\sum_1^n{}_c$ implies that the sum is over all *independent* combinations of the digits 1 to n for the subscripts.

For example,

$$a_1^{(r)} = (-1) \sum_{k=1}^n \alpha_k^{2^r}.$$

On the basis of (3), it can be shown that, if

$$|\alpha_1| > |\alpha_2| > |\alpha_3| > \cdots > |\alpha_n|,$$

then

$$\lim_{r \to \infty} \left| \frac{a_j^{(r)}}{a_{j-1}^{(r)}} \right|^{1/2^r} = |\alpha_j|$$

with $a_0^{(r)} = 1$, where the values of $a_j^{(r)}$, for any r can be calculated from (2).

The signs of the α_j can be determined from either rough graphs or substitution in $P_n(x)$.

This method can be extended to polynomials with equal and/or complex roots {see Bareiss [7], [8]}.

PROBLEMS

Determine graphically the mentioned roots of the following equations:

1. $x^2 - \cos \pi x = 0, \; x > 0.$

2. $x - \cos^2 \pi x = 0,$ all roots.

3. $(x-1)^2 - \frac{1}{2} e^x = 0,$ all roots.

4. $(x-1)^2 - e^{-x} = 0, \; x > 0.$

5. $e^{-x} - 2 + x^2 = 0.$

6. $2\sqrt{x} - \cos\frac{\pi x}{2} = 0,$ all roots.

7. $\sqrt{x+1} - \frac{1}{x} = 0,$ all roots.

8. $\sqrt{x} - 2\cos\frac{\pi x}{2} = 0,$ the smallest positive root.

9. $3x - \tan\frac{\pi x}{4} = 0,$ the smallest positive root.

10. $x^2 - \cot\frac{\pi x}{3} = 0,$ the smallest positive root.

11. $x^2 - \cos^2 \pi x = 0,\ x > 0.$

12. $x^2 - \sin \pi x = 0,$ all roots.

Determine to an accuracy of 10^{-5} all the roots of the following equations using any one of the methods **1–3**:

13. $x - \sin x = 0.25.$

14. $x^2 - 20 \sin x = 0.$

15. $x^2 - \sin 5x = 0.$

16. $1.8x^2 - \sin 10x = 0.$

17. $x^2 - \sin \pi x = 0.$

18. $x - \cos x = 0.$

19. $x^2 - \cos \pi x = 0.$

20. $x - \cos^2 \pi x = 0.$

21. $x^2 - \cos^2 \pi x = 0.$

22. $2\sqrt{x} - \cos\frac{\pi}{2}x = 0.$

23. $\sqrt{x} - 2\cos\frac{\pi}{2}x = 0.$

24. $\sqrt{x} - \cos(0.387x) = 0.$

25. $x - 3\cos^2(1.04x) = 0.$

26. $x - \cos\left(\frac{0.7854 - x\sqrt{1+x^2}}{1+2x^2}\right) = 0.$

27. $2.2x - 2^x = 0.$

28. $2^x - 2x^2 - 1 = 0.$

29. $2^{x-1} + 4^{x-2} - 8^{x-2} - 0.2x = 0.$

30. $2^x - 4x = 0.$

31. $x^x + 2x = 6.$

32. $2 \ln x - \frac{1}{x} = 0.$

33. $2 \log x - \frac{x}{2} + 1 = 0.$

34. $\log x - \frac{1}{x^2} = 0.$

35. $\log x - 7/(2x + 6) = 0.$

36. $x \log x - 1.2 = 0.$

37. $\log (3x - 1) + e^{2x-1} = 0.$

38. $e^{-x} + x^2 - 2 = 0.$

39. $e^{-x} - (x - 1)^2 = 0.$

40. $e^x + x^2 - 2 = 0.$

41. $e^x - 2(x - 1)^2 = 0.$

42. $2 - x e^x = 0.$

43. $x - a^x = 0,\ a = 0.5\,(0.1)\,1.4.$*

44. $x e^{-x} - e^{-a-1} = 0,\ a = 1\,(0.1)\,3.4.$

Determine to an accuracy of 10^{-5} the mentioned roots of the following equations using any one of the methods **1–3**:

45. $x^2 - \cos x = 0,\ x < 0.$

46. $\frac{1}{x} - \pi \cos \pi x = 0,$ the smallest positive root.

47. $\sin x - 0.6\, x \cos x = 1,$ the first two positive roots.

48. $\sin x - \cos^2 x + 0.25 = 0,$ the first two positive roots.

49. $\sec x - x^2 - 1 = 0,$ the smallest positive root.

50. $x - e^{\pi \cot x} = 0,$ the smallest positive root.

51. $\cot 1.05\, x - x^2 = 0,$ the smallest positive root.

52. $\cot x + x/(1 - x^2) = 0,$ the smallest positive root.

53. $x^2 - \cot \frac{\pi x}{3} = 0,$ the smallest positive root.

54. $\tan x - x = 0,$ the smallest positive root.

*) See the Editor's Note for an explanation of this notation: $a = 0.5\,(0.1)\,1.4.$

55. $\tan \frac{\pi x}{4} - x - 3 = 0,$ the largest negative and the smallest positive roots.

56. $3x - \tan \frac{\pi x}{4} = 0,$ the smallest positive root.

57. $6x - 5 \sinh x = 0, \; x < 0.$

58. $\tan x + \tanh x = 0,$ the smallest positive root.

59. $x \sinh \frac{10}{x} - 15 = 0, \; x > 0.$

60. $\cosh x \cdot \cos x = 1,$ the smallest positive root.

61. $\cosh x \cdot \cos x = -1,$ the smallest positive root.

62. $1 + x - \exp(3x/4) = 0, \; x \neq 0.$

63. $3x - e^x = 0,$ the greatest positive root.

64. $e^x + e^{-3x} - 4 = 0, \; x < 0.$

65. $2x - \log x - 7 = 0,$ the greatest positive root.

66. $5x - 8 \ln x - 8 = 0,$ the smallest positive root.

67. $\cot x - \frac{x}{a} = 0, \; a = 1\,(1)\,25,$ the smallest positive root.

68. $1 - x^{-a} + (1 - x)^{-a} = 0, \; a = 0.5\,(0.1)\,2.8,$ the smallest positive root.

69. $x^{4n} + 4n x^{2n} - 1 = 0, \; n = 1\,(1)\,4, \; x > 0.$

70. $\cot x - \frac{1}{\pi} \ln a = 0, \; a = 1.5\,(0.05)\,2.7,$ the smallest positive root.

71. $\tan x + a x/(x^2 - a) = 0, \; a = 1\,(0.1)\,1.9, \; x \in (\pi/2, \pi).$

72. $\tan(a x + b) = x^2, \; x \in (0, 1.5).$

Table of values for a and b

b	a					
0.1	0.58	0.56	0.55	0.53	0.52	0.50
0.2	0.56	0.52	0.50	0.47	0.46	0.44
0.3	0.48	0.46	0.44	0.42	0.40	0.37
0.4	0.40	0.38	0.36	0.33	0.31	0.30

Using **4**, determine all the roots of the following equations to an accuracy of 10^{-5}:

73. $\ln x + (x + 1)^3 = 0.$

74. $x \cdot 2^x = 1.$

75. $\sqrt{x + 1} = \frac{1}{x}.$

76. $x - \cos x = 0.$

77. $3x - \cos x - 1 = 0.$

78. $2\sqrt{x} = \cos \frac{\pi x}{2}.$

79. $x - \sin x = 0.25.$

80. $x - 0.21 \sin (0.5 + x) = 0.$

81. $x + \lg x = 0.5.$

82. $x \lg x = -0.125.$

83. $x \ln x = 100.$

84. $2 - x = \ln x.$

85. $(x - 1)^2 - \frac{1}{2} e^x = 0.$

86. $2 - x e^x = 0.$

87. $(4 + x^2)(e^x - e^{-x}) = 18.$

Using **4**, determine the mentioned roots of the following equations to an accuracy of 10^{-5}:

88. $x^3 - x - 1 = 0, \ x \in (1.3, \ 1.4).$

89. $x^3 + x - 1000 = 0,$ the greatest positive root.

90. $x^3 - 4x^2 + 10x - 10 = 0, \ x \in (1.5, \ 1.7).$

91. $2.2x - 2^x = 0,$ the smallest root.

92. $x^2 + 4 \sin x = 0, \ x \neq 0.$

93. $x^2 - \cos x = 0, \ x > 0.$

94. $x \tan x = 1.28,$ the smallest positive root.

95. $6x - 5 \sinh x = 0, \ x > 0.$

96. $3 \cosh x - \sinh x - 9 = 0, \ x > 0.$

97. $2x - \log x = 7,$ the greatest positive root.

98. $5x - 8 \ln x = 8,$ the greatest positive root.

99. $3x - e^x = 0,$ the smallest positive root.

100. $e^{-x} - 2 + x^2 = 0, \ x > 0.$

101. $(x - 1)^2 - e^{-x} = 0, \ x \neq 0.$

102. $e^x + e^{-3x} = 4, \ x > 0.$

103. $x + e^x + e^{-3x} - 4 = 0, \ x > 0.$

Determine the roots of the polynomial

$$x^3 + a x^2 + b x + c = 0$$

using Lobachevskiy's method, and improve their accuracy using Newton's method. For the coefficients a, b and c given in the following table, determine all the roots to six significant places.

Table of coefficients a, b and c

Problem number	a	b	c
104	−14.4621	60.6959	−70.9238
105	−10.2374	−91.2105	492.560
106	−19.7997	28.9378	562.833
107	−5.57496	−193.022	−633.105
108	23.7997	44.9378	−703.378
109	9.57496	−243.672	773.651
110	20.2374	−131.210	−843.923
111	38.4621	364.594	914.196
112	−13.3667	39.8645	−20.6282
113	2.65804	−28.0640	21.9032
114	−6.49510	−31.2543	23.1782
115	9.92960	17.8390	−24.4532
116	6.09510	−35.3942	−25.7283
117	−10.3296	20.3641	27.0033
118	−1.65804	−34.3767	−28.2783
119	14.7667	54.3542	29.5533
120	−14.9667	56.4241	−30.8283
121	−5.09510	−45.7440	32.1033
122	−12.9621	50.7295	−57.8098
123	−4.49967	−24.4975	59.5583
124	9.31268	7.47465	−61.3069

Using Lobachevskiy's method, determine the roots of the polynomial

$$a\,x^4 + b\,x^3 + c\,x^2 + dx + e = 0$$

to four significant places. Using Newton's method, improve their accuracy to six significant places. The values for the coefficients a, b, c, d and e are given in the following table.

Table of coefficients a, b, c, d and e

Problem number	a	b	c	d	e
125	7.5046	11.0866	3.8239	0.4420	−0.4797
126	10.0128	26.6918	23.3954	20.1112	7.1068
127	4.0716	12.7356	8.7042	10.8110	−18.8742
128	0.1357	1.3038	1.3773	−2.8696	20.6448

Problem number	a	b	c	d	e
129	2.9127	1.4726	12.5882	27.4317	12.6488
130	1.0098	4.5603	10.9752	3.0800	−10.2935
131	0.6786	2.7432	−10.3531	35.3539	100.6647
132	0.1254	−0.0027	−4.7285	−25.0263	−33.0434
133	8.9958	−15.6291	7.0606	−1.4688	0.3111
134	8.7354	0.1021	−4.3251	−3.6443	−0.6985
135	3.2004	−0.7934	−16.3394	−26.4327	−11.8793
136	1.7125	−3.4192	−10.0420	−16.6985	28.1418
137	0.1224	−0.1741	−8.0008	−29.0426	−42.4941
138	0.0253	−0.5224	2.9153	−7.4376	15.1869
139	0.3649	3.8105	9.4893	9.0549	−88.0308
140	9.4481	5.0141	−4.3670	−0.2807	−0.3477
141	9.2214	−8.9079	7.7486	−1.3684	−0.9893
142	2.4605	15.5686	35.5695	39.0466	14.8393
143	2.9651	4.8376	−16.4157	41.1832	−33.1773
144	0.2247	3.6517	16.8013	33.7320	63.8746
145	10.2543	−5.8976	−3.9975	0.4890	−0.3519
146	5.5969	−39.2424	76.9317	−75.2196	53.9069
147	1.9402	−7.9076	5.3741	−11.6349	−28.1685
148	0.1445	−0.8262	−5.8774	−13.4301	−63.0958
149	0.7708	0.1560	7.2999	−9.6481	−15.6894
150	0.5652	−4.1834	17.2128	−30.9053	17.2894
151	9.5457	15.4606	7.5485	1.6612	0.3431
152	1.5846	2.5970	−11.5105	7.0066	−14.8092
153	7.2065	−2.8084	11.1603	7.3542	−4.2921
154	2.2704	4.6736	−14.2683	14.4069	33.1344
155	1.5142	1.7629	−12.0626	57.4659	−57.1228
156	0.1073	2.0509	12.3213	35.0630	74.2672
157	0.2397	0.4987	−29.4663	146.1143	−316.7604
158	0.6634	−0.4725	4.2312	17.7903	12.9492
159	0.5069	−1.9084	3.3582	12.4847	−14.1824
160	5.9013	−14.4611	4.1920	1.5088	2.9848
161	1.5604	−16.5042	47.3564	−62.4360	61.4922
162	3.2395	7.6929	3.1983	−5.4840	−2.2204
163	4.8608	0.4207	11.6158	−27.7636	10.7115
164	1.0864	−8.5604	17.3658	1.1890	−46.9558
165	0.8648	−11.2956	51.6184	−103.7830	72.6754
166	0.1596	−0.8671	−16.0488	−44.1193	−174.2580
167	0.5208	2.8211	3.3012	−14.9702	−15.8556
168	6.7671	2.7746	−16.3020	9.9877	−3.3971
169	3.2129	16.2999	9.1360	−8.1953	24.2977
170	1.5522	3.0668	−29.0177	32.8239	−61.3827

Problem number	*a*	*b*	*c*	*d*	*e*
171	8.3486	6.0411	16.8350	18.9312	5.1617
172	2.4582	4.9326	10.5595	3.3620	−5.1516
173	0.4539	1.6457	−1.6669	13.7774	21.5066
174	0.6786	1.7511	−6.9269	43.5545	−58.2978
175	0.1176	1.8851	−1.6427	−28.4306	127.0896
176	0.4879	3.0956	13.2783	26.0996	15.1655
177	9.1260	−7.2930	−17.8207	−6.1219	−4.4187
178	9.1904	−31.4158	34.9350	−25.0912	6.7564
179	0.5858	−0.3803	−13.5386	−27.9452	−23.0287
180	4.7443	12.4447	7.0642	−17.4259	−9.5033
181	2.3331	1.5172	6.4455	−24.6476	12.3891
182	0.1106	−0.8637	1.6530	−1.4697	−9.2106
183	0.2449	−3.7522	19.6856	−45.2855	35.9808
184	0.1224	−3.8778	40.7718	−178.8159	388.9881
185	0.3619	2.5021	0.4707	−5.1729	15.2334
186	0.1972	1.1641	−4.4449	−7.9605	−22.2024
187	3.3369	−0.4442	−2.5872	24.0483	6.6044
188	0.8635	−2.4440	2.3387	5.4038	−3.8564
189	0.2484	2.1115	6.3911	17.5422	18.0486
190	0.6006	0.0681	−16.4096	75.2080	−86.6617
191	0.0806	2.1478	17.5999	64.4501	202.8204
192	9.8892	−5.3727	−1.0537	−1.4451	0.6943
193	7.8387	−20.8815	12.9362	−3.1591	−7.0295
194	0.2632	−1.5478	−5.8590	10.3994	−37.4518
195	1.4014	1.6516	5.3095	−7.0659	−6.2986
196	1.6592	0.9760	6.7519	−29.7247	17.1084
197	0.7371	−2.5705	−6.7880	−62.9515	−108.8661
198	0.1545	−2.5917	15.0014	−38.2283	34.4641
199	0.1029	−0.5810	−17.7223	−80.2035	−388.8245
200	0.1358	−1.8240	9.9088	−37.5808	56.0056
201	0.4403	−1.3263	0.1958	22.6079	22.5645
202	0.1155	−1.7454	6.5405	−0.1852	−71.0436
203	0.0455	0.2733	−14.7739	97.3584	−395.4034
204	0.2373	0.8973	1.2113	−31.6250	40.6756
205	0.0333	0.5686	1.8581	0.9696	−57.7257
206	0.0741	0.2943	−7.0199	93.3695	−300.0184
207	0.0979	−1.5828	10.2121	−4.5966	−82.0787
208	0.5133	1.8130	5.8727	−18.1846	−25.6619
209	0.1222	1.5317	4.6530	3.5228	−71.3707
210	0.0638	−1.6329	1.4173	89.0564	589.7794
211	0.0287	−0.4350	1.7986	−25.1499	79.2396
212	0.1526	0.4224	6.1438	36.7596	43.8816

Problem number	a	b	c	d	e
213	0.0686	1.1998	7.2357	1.8159	−56.5551
214	0.0637	−1.2036	8.3031	−30.6430	40.5411
215	0.1482	−0.2916	0.6889	10.2143	10.8433
216	0.0231	0.0618	−11.5155	115.7682	−421.2744
217	0.1482	−1.3583	8.8663	−7.3860	−46.3068
218	0.2299	−1.0736	2.1774	70.1089	117.3524
219	0.1034	−0.9995	4.9207	−9.9957	6.3162
220	0.0494	1.6499	15.8935	70.2523	246.8088
221	0.0588	0.2282	0.5214	−2.2962	−3.3282
222	0.0621	0.9654	3.4867	4.3866	−71.3538
223	0.2074	−1.4001	5.6687	28.4368	−59.9872
224	0.0558	−0.8930	−6.2911	49.2607	−271.1385
225	0.2945	−1.5086	3.7762	13.9522	−21.0103
226	0.1067	−2.4968	20.8671	−89.9776	137.1019
227	0.0657	0.7128	4.9757	14.5174	12.5184
228	0.2034	−2.7149	18.9265	−57.6003	53.5869
229	0.1554	−0.9935	5.4431	28.1693	−71.5427
230	0.1365	−2.0951	7.9800	3.9354	−56.5488
231	0.0301	0.4893	−3.3234	−27.6864	−156.8416
232	0.1085	−1.1086	5.7172	−12.7049	8.5358
233	0.2679	−1.2865	8.9964	−9.6751	−40.3189
234	0.0551	0.7235	1.2115	38.5702	136.3638
235	0.0749	0.5485	−0.4292	−21.4071	38.8773
236	0.2256	2.9187	9.3939	4.9872	−93.3012
237	0.0377	−1.3916	16.6331	−83.9940	200.3436
238	0.0513	0.9852	7.0142	27.1607	36.9054
239	0.0287	−0.1156	−8.6341	−71.7148	−294.9705
240	0.2704	1.5782	10.8886	10.5070	−47.0162
241	0.0343	−0.1050	−2.7012	−28.1439	−75.1132
242	0.0671	−1.4562	7.9356	−11.9900	−280.2804

Determine all the roots of the following equations to five significant places:

243. $5x^3 + 2x^2 - 15x - 6 = 0.$

244. $x^3 + x^2 - 10x + 8 = 0.$

245. $x^3 - 4x^2 + 2 = 0.$

246. $x^3 + 3x^2 - 3 = 0.$

247. $x^3 - 3x^2 - 3 = 0.$

248. $x^3 - x - 1 = 0.$

249. $x^3 - 7x - 7 = 0.$

250. $x^3 - 8x + 15 = 0.$

251. $x^4 - 6x^3 + 11x^2 + 2x - 28 = 0.$

252. $x^4 - 3x^3 + 8x^2 - 5 = 0.$

253. $x^4 - 3x^2 + 4x - 1 = 0.$

254. $x^4 + 7.64x^3 + 23.604x^2 + 38.910x + 38.149 = 0.$

255. $1.23x^5 - 2.52x^4 - 16.1x^3 + 17.3x^2 + 29.4x - 1.34 = 0.$

256. $7.5x^5 + 5.44x^3 - 3.24x^2 - 1.85x + 0.2 = 0.$

257. $x^5 + x^4 - x^3 - 0.2x^2 + 3x + 0.5 = 0.$

258. $1.5x^5 - 3x^4 + 7x^3 + 0.5x^2 - 2x - 2 = 0.$

259. $x^5 - 0.27x^4 - 1.93x^3 + 7.27x^2 - 0.37x + 1.01 = 0.$

260. $x^5 + 13x^4 + 27x^3 + 42x^2 - 27x + 19 = 0.$

261. $2x^5 + x^4 + 2x^3 - x^2 + 2x - 1 = 0.$

262. $x^5 + 0.6x^4 + 0.7x^3 + 0.8x^2 + 0.9x + 1 = 0.$

263. $x^5 - x^4 + 2x^3 - 3x^2 + 4x - 5 = 0.$

264. $x^5 + 2.24x^4 - 1.12x^3 - 0.93x^2 + x + 1 = 0.$

265. $x^5 + x^4 + 1.1x^3 - 3x^2 + x + 1.7 = 0.$

266. $x^5 + 7x^4 - 3x^3 + 2x^2 - 0.5x + 3 = 0.$

267. $x^5 - x^4 + 5x^3 - x^2 + 2x + 2 = 0.$

268. $x^5 - 7.9x^4 + 24.46x^3 - 37.074x^2 + 27.512x - 8.0042 = 0.$

269. $x^6 - 2x^5 + 1.7x^4 - 4.72x^3 - 0.86x^2 + 4.62x - 7 = 0.$

270. $x^6 - 7x^5 + 3x^4 + x^3 - x^2 + 0.5x - 2 = 0.$

271. $3.26x^6 + 4.2x^4 + 3.08x^3 - 7.16x^2 + 1.92x - 7.76 = 0.$

272. $x^6 - 6x^5 + 3x^4 + 5x^3 - 6x + 2 = 0.$

273. $x^6 - 3x^2 + x - 1 = 0.$

274. $x^6 + 4.2240x^5 + 6.5071x^4 + 7.5013x^3 + 8.4691x^2 +$
$+ 3.3641x + 1.6252 = 0.$

275. $x^7 + 0.456x^6 - 0.427x^5 + 0.472x^4 - 0.482x^3 + 0.186x^2 -$
$- 0{,}767x + 0.205 = 0.$

276. $x^7 - 2x^5 - 3x^3 + 4x^2 - 5x + 6 = 0.$

277. $x^7 - 3.5x^6 + 4.8462x^5 - 3.3654x^4 + 1.2238x^3 - 0.2203x^2 +$
$+ 0.016317x - 0.0002914 = 0.$

278. $x^7 - x^6 + 8x^5 - 4x^4 + 6x^3 + 2x^2 - 5x + 1 = 0.$

279. $x^7 + 0.79\,x^6 - 0.35\,x^5 + 0.36\,x^4 - 3.87\,x^3 - 1.05\,x^2 + 0.66\,x + 1 = 0.$

280. $x^8 + 5\,x^7 + 3\,x^5 - 2\,x^4 - 7\,x^3 + 9\,x - 10 = 0.$

281. $128\,x^8 - 256\,x^6 + 160\,x^4 - 32\,x^2 + 1 = 0.$

282. $x^8 + 3.42\,x^7 - 7.06\,x^6 + 0.34\,x^5 - 6.43\,x^4 + 0.98\,x^3 + 3.49\,x^2 - 9.88\,x + 5.55 = 0.$

283. $x^8 - 5\,x^7 + 3\,x^6 + 8\,x^5 - 35\,x^4 + 73\,x^3 - 6x^2 + 23x + 7 = 0.$

284. $0.679\,x^8 - 0.137\,x^7 + 0.098\,x^6 + 0.543\,x^5 - 0.586\,x^4 - 0.037\,x^3 + 0.906\,x^2 - 0.345\,x + 0.777 = 0.$

285. $x^9 + 3\,x^8 - x^7 + x^6 + 6\,x^5 - 7\,x^4 + x^3 + x^2 - x + 2 = 0.$

286. $x^9 + 34\,x^8 - 2\,x^7 + 24\,x^6 - 76\,x^5 + 33\,x^4 - x^3 + 3\,x^2 + 7\,x - 33 = 0.$

287. $512\,x^{10} - 1280\,x^8 - 1120\,x^6 - 400\,x^4 + 50\,x^2 - 1 = 0.$

288. $2048\,x^{12} - 6144\,x^{10} + 6912\,x^8 - 3584\,x^6 + 840\,x^4 - 72\,x^2 + 1 = 0.$

2. The solution of systems of non-linear equations

THEORY

7. Newton's method for two simultaneous equations {see [3; § 8.8]}

A system of two simultaneous equations can be written in the form

$$\begin{cases} F(x, y) = 0, \\ G(x, y) = 0. \end{cases}$$

For this alignment, the method of Newton becomes:

$$x_{n+1} = x_n - \frac{1}{I(x_n, y_n)} \begin{vmatrix} F(x_n, y_n) & F'_y(x_n, y_n) \\ G(x_n, y_n) & G'_y(x_n, y_n) \end{vmatrix},$$

$$y_{n+1} = y_n - \frac{1}{I(x_n, y_n)} \begin{vmatrix} F'_x(x_n, y_n) & F(x_n, y_n) \\ G'_x(x_n, y_n) & G(x_n, y_n) \end{vmatrix},$$

where

$$I(x, y) = \begin{vmatrix} F'_x(x, y) & F'_y(x, y) \\ G'_x(x, y) & G'_y(x, y) \end{vmatrix} \neq 0 \qquad (n = 0, 1, 2, \ldots).$$

8. An iteration method for systems of equations {see [4; chapter 3]}

The system of equations

$$F_i(x_1, x_2, \ldots, x_k) = 0 \qquad (i = 1, 2, \ldots, k)$$

can be rewritten as

$$x_i = \varphi_i(x_1, x_2, \ldots, x_k).$$

The successive approximations $x_i^{(n)}$ $(n = 1, 2, \ldots)$ are then defined by the iteration process

$$x_i^{(n)} = \varphi_i(x_1^{(n-1)}, x_2^{(n-1)}, \ldots, x_k^{(n-1)}).$$

If this process converges, i.e.,

$$\lim_{n\to\infty} x_i^{(n)} = \alpha_i,$$

then the α_i define a root of the original system.
A suitable initial approximation must be determined.

PROBLEMS

Solve, to five significant places, the following systems of equations using Newton's method. The initial approximations should be determined graphically.

289. $\begin{cases} \sin(x + y) - a x = k, \\ x^2 + y^2 = 1, \ x > 0, \ y > 0, \end{cases}$
$a = 1.1\,(0.1)\,1.6, \ k = -0.2\,(0.1)\,0.2.$

290. $\begin{cases} \tan(x y + k) = x^2, \\ a x^2 + 2 y^2 = 1, \ x > 0, \ y > 0, \end{cases}$
$a = 0.5\,(0.1)\,1.0, \ k = 0.0\,(0.1)\,0.4.$

291. $\begin{cases} e^{xy} = x^2 - y + a, \\ (x + 0.5)^2 + y^2 = k, \ x > 0, \ y > 0; \end{cases}$
$a = 1.0\,(0.1)\,1.5, \ k = 0.6\,(0.1)\,1.0,$

292. $\begin{cases} e^{x^2 - 1 - y} = x^2 + (y + a)^2 - 0.4, \\ x^2 - y^2 = k, \ x > 0, \ y > 0, \end{cases}$
$a = 0.0\,(0.1)\,0.5, \ k = 0.1\,(0.1)\,0.5.$

Solve, to five significant places, the following systems of equations:

293. $\begin{cases} 2 x^2 - x y - 5 x + 1 = 0, \\ x + 3 \log x - y^2 = 0. \end{cases}$

294. $\begin{cases}(x - 1.2)^2 + (y - 0.6)^2 = 1, \\ 4.2\,x^2 + 8.8\,y^2 = 1.42.\end{cases}$

295. $\begin{cases}x^2 y^2 - 3\,x^3 - 6\,y^3 + 8 = 0, \\ x^4 - 9\,y + 2 = 0.\end{cases}$

296. $\begin{cases}\sin x - y = 1.32, \\ \cos y - x = -0.85.\end{cases}$

297. $\begin{cases}x^7 - 5\,x^2 y^4 + 1510 = 0, \\ y^3 - 3\,x^4 y - 105 = 0.\end{cases}$

298. $\begin{cases}40\,[\sin(4\,\pi\,x)\sin(2\,\pi\,y) - \sin(2\,\pi\,x)\sin(4\,\pi\,y)] = -30, \\ 40\,[\cos(8\,\pi\,x) + \cos(8\,\pi\,y)] = 10.\end{cases}$

Let R denote the region bounded externally by the straight lines

$$y = 0,\ y = x,\ x = 0.5.$$

In R, determine the roots of the following systems of equations:

299. $\begin{cases}2\,x^2 - x\,y - y^2 + 2\,x - 2\,y + 6 = 0, \\ y - x - 1 = 0.\end{cases}$

300. $\begin{cases}x = \log(y/z) + 1, \\ y = 0.4 + z^2 - 2\,x^2, \\ z = 2 + x\,y/20,\ x_0 = 1,\ y_0 = 2.2,\ z_0 = 2.\end{cases}$

301. $\begin{cases}a\,x^3 - y^2 - 1 = 0, \\ x\,y^3 - y - 4 = 0,\ a = 1\,(0.5)\,7.5.\end{cases}$

REFERENCES

[1] B. P. Demidovich and I. A. Maron, *Basic Numerical Mathematics* (russian), Chapters 4 and 5, Nauka, Moscow, 1970.

[2] C.-E. Fröberg, *Introduction to Numerical Analysis,* Addison – Wesley, Reading, Mass., 1966.

[3] A. Ralston, *A First Course in Numerical Analysis,* McGraw-Hill, New York, 1965.

[4] D. G. Moursund and C. S. Duris, *Elementary Theory and Application of Numerical Analysis,* McGraw-Hill, New York, 1967.

[5] M. L. James, G. M. Smith and J. C. Wolford, *Applied Numerical Methods,* International Textbook Company, Scranton, Penn., 1967.

[6] I. S. Berezin and I. A. Zhidkov, *Computing Methods* (original in Russian), Volume II, Pergamon Press, Oxford, 1965.

[7] E. H. Bareiss, *Resultant procedure and the mechanization of the Graeffe process,* J. Assoc. Comp. Mach. 7 (1960), 346–386.

[8] E. H. Bareiss, *The root-cubing and the general rootpowering methods for finding the zeros of polynomials,* Argonne National Laboratory Report *ANL*–7344, 1967.

Chapter 2

NUMERICAL METHODS IN LINEAR ALGEBRA *)

Consider the following system of n (linear algebraic) equations in n unknowns

$$\left.\begin{array}{l} a_{11}\, x_1 + a_{12}\, x_2 + \cdots + a_{1n}\, x_n = b_1, \\ a_{21}\, x_1 + a_{22}\, x_2 + \cdots + a_{2n}\, x_n = b_2, \\ \cdot\ \cdot\ \cdot\ \cdot\ \cdot\ \cdot\ \cdot\ \cdot\ \cdot\ \cdot\ \cdot\ \cdot \\ a_{n1}\, x_1 + a_{n2}\, x_2 + \cdots + a_{nn}\, x_n = b_n. \end{array}\right\} \tag{a}$$

We introduce the following notation: the matrix of coefficients of (a) is denoted by $A = [a_{ij}]$, the column of right hand sides by b and the column of unknowns by x, i.e.,

$$A = \begin{bmatrix} a_{11}\, a_{12} \cdots a_{1n} \\ a_{21}\, a_{22} \cdots a_{2n} \\ \cdot\ \ \cdot\ \ \cdot\ \ \cdot\ \ \cdot \\ a_{n1}\ a_{n2} \cdots a_{nn} \end{bmatrix}, \quad b = \begin{bmatrix} b_1 \\ b_2 \\ \vdots \\ b_n \end{bmatrix}, \quad x = \begin{bmatrix} x_1 \\ x_2 \\ \vdots \\ x_n \end{bmatrix}.$$

Thus, using this notation, (a) becomes

$$A\, x = b.$$

We denote the *determinant* of A by $\det A$, the *transpose* by A^T {i.e., if $A = [a_{ij}]$, then $A^T = [a_{ji}]$} and the *inverse* by A^{-1} {i.e., $A^{-1} \cdot A = A \cdot A^{-1} = I$, where I is the unit matrix}.

If $\det A \neq 0$, A is *non-singular* and (a) has a unique solution, which can be written symbolically as

$$x = A^{-1}\, b.$$

If $b = 0$, (a) is said to be *homogeneous*. A homogeneous system has a non-zero solution if, and only if, $\det A = 0$.

*) For a more detailed discussion of the elementary notation and results of linear algebra, see [1].

With respect to the matrix A, we can define the *characteristic equation*

$$|A - \lambda I| = \begin{vmatrix} a_{11} - \lambda & a_{12} & \cdots a_{1n} \\ a_{21} & a_{22} - \lambda & \cdots a_{2n} \\ \cdot & \cdot & \cdot \\ a_{n1} & a_{n2} & \cdots a_{nn} - \lambda \end{vmatrix} \tag{b}$$

$$= (-1)^n (\lambda^n - p_1 \lambda^{n-1} - p_2 \lambda^{n-2} - \cdots - p_{n-1} \lambda - p_n) = 0.$$

The polynomial in λ is called the *characteristic polynomial.* The roots, λ_i $(i = 1, 2, \ldots, n)$, of (b) are called the *eigenvalues* {latent roots, characteristic roots, proper values or proper numbers} of A. Corresponding to each eigenvalue λ_i, there exists an *eigenvector* (latent, characteristic or proper vector) x_i which is a non-trivial {non-zero} solution of

$$(A - \lambda_i I)\, x_i = 0. \tag{c}$$

Thus, it is necessary to solve (c) n times in order to determine all the eigenvectors of A. At is clear that, except for a constant of proportionality, the components of the eigenvector x_i are unique. The number of linearly independent eigenvectors associated with one and the same eigenvalue does not exceed the multiplicity of that eigenvalue. Hence, if all the eigenvalues are distinct, then there exists, upto a constant of proportionality, one and only one eigenvector corresponding to each eigenvalue.

The solution of linear systems of the form (a), the inversion of matrices A and the determination of its eigenvalues and eigenvectors are the subjects of this chapter.

1. The solutions of systems of linear equations*)

THEORY

As a consequence of using floating point arithmetic with a finite mantissa [2; chapter 20], and thus, the introduction of rounding errors at intermediate stages throughout a calculation, the use of even exact methods to solve (a) yields only approximate (and not the exact) solutions.

*) The ideal reference is Forsythe and Moller [2]. See also, [3], [4], [5] and [6].

The accuracy of the solution depends essentially on the *condition* of A. A matrix A is said to be well-conditioned if small relative changes in its coefficients and right hand sides leads to insignificant relative changes in the solution.
The *condition* of A is defined in terms of a *condition number*. Two examples are:
Turing's number

$$M\text{-number} = n \cdot \max_{ij} |a_{ij}| \max_{ij} |d_{ij}|,$$

where d_{ij} is an element of the inverse matrix;
Todd's number

$$P\text{-number} = \frac{\max |\lambda_i|}{\min |\lambda_i|},$$

where λ_i is an eigenvalue of the matrix A.
If the condition number is large, the system is said to be *ill-conditioned.*
Methods for solving systems of linear equations can be divided into two basically distinct groups:
1) exact, and 2) iterative. In exact methods, the solution is determined after a finite number of arithmetic operations which are specified in advance. In iterative methods, the solution can be evaluated to practically any degree of accuracy by means of convergent infinite processes.
Below, computational details are given for some of the methods which can be used to solve (a).

Exact methods

Methods 1 and 2 are elimination methods (see [2; chapter 9], and [3; chapter 4]): the initial system (a) is reduced to upper triangular form which defines a simple computational scheme for the determination of the unknown x_i.

1. Gaussian elimination*) with complete pivoting

Initially, the largest element (*pivotal element*) in modulus of A is selected. For example, let it be a_{pq}. Row p containing this pivotal

*) *Editor's Note.* In modern English texts, for example [2] and [4], this is more usually referred to as *LU* factorization.

element is called the *pivotal row*. For all $i \neq p$, the multiplyers

$$m_i = \frac{a_{iq}}{a_{pq}} \tag{22.1}$$

are calculated. Then the following transformation is made to the augmented matrix of (a): for each i, the elements of the pivotal row are multiplied by m_i and subtracted term by term from the terms of row i; then row p and column q are deleted from this transformed augmented matrix in order to obtain an auxiliary matrix. The operations performed on the original augmented matrix are applied to on this auxiliary matrix. The procedure continues inductively until a matrix with one row of two elements is obtained – the final pivotal row.

Recombining all the pivotal rows with suitable rearranging (permutation), an upper triangular system is obtained which is equivalent to (a) (see [4; chapter 4, § 39]). This stage of the computation is called the *forward elimination*.

Solving the upper triangular system starting with the final pivotal row, we obtain the values of the unknown x_i. This stage of the computation is called the *back substitution*.

Note. This method can be used to calculate the value of det (A). It equals the product of the pivotal elements with the correct sign, which depends on the rearrangement into upper triangular form, attached (see [4; chapter 4, § 41] and [2; chapter 14]).

2. Compact Gaussian elimination without pivoting*)

In this method A is transferred into the product of a lower triangular matrix B and an upper triangular matrix C

$$A = BC,$$

with one of them, for example C, having only unit, diagonal elements, so that (a) is equivalent to the system with coefficient matrix C.

Since the same operations are performed on the right hand sides as on the coefficients, for compactness of description A will now denote the augmented matrix of (a) and C will denote the aug-

*) *Editor's Note*: Sometimes referred to as the method of Crout or Gauss-Doolittle.

mented matrix of the transformed system. Then, the process of determining B and C is defined inductively by

$$b_{ij}^{(2j-1)} = a_{ij}^{(0)} - \sum_{k=1}^{j-1} b_{ik}^{(2k-1)} c_{kj}^{(2k)} \qquad (i = j, j+1, \ldots, n),$$

$$c_{jm}^{(2j)} = \frac{a_{jm} - \sum_{k=1}^{j-1} b_{jk}^{(2k-1)} c_{km}^{(2k)}}{b_{jj}^{(2j-1)}} \qquad (m = j+1, j+2, \ldots, n+1).$$

The values of the unknowns $x_1, x_2, \ldots, x_n$ are then found successively by *back substitution*

$$x_i + \sum_{k=i+1}^{n} c_{ik}^{(2i)} x_k = c_{i\,n+1}^{(2i)} \qquad (i = n, n-1, \ldots, 1).$$

This compact scheme is applicable only when the determinants of the principal minors of A are non-singular, i.e., when

$$a_{11}, \begin{vmatrix} a_{11} & a_{12} \\ a_{21} & a_{22} \end{vmatrix}, \ldots, \det A \neq 0.$$

Note. The factorization can be used to invert A. If we denote the elements of A^{-1} by $[A^{-1}]_{ij} = d_{ij}$, then the following systems of equations can be used to determine them:

$$\begin{cases} \qquad\qquad i = 1, 2, 3, \ldots, n, \\ \sum_{j=1}^{n} d_{ij} b_{j1}^{(1)} = 1 \quad 0 \quad 0 \cdots 0, \\ \sum_{j=2}^{n} d_{ij} b_{j2}^{(3)} = \quad\; 1 \quad 0 \cdots 0, \\ \cdots\cdots\cdots\cdots \\ d_{nn} b_{nn}^{(2n-1)} = \qquad\qquad 1 \end{cases}$$

$$\begin{cases} \qquad\qquad j = 2, 3, \ldots, n, \\ d_{1j} + \sum_{i=2}^{n} d_{ij} c_{1i}^{(2)} = 0 \quad 0 \cdots 0, \\ d_{2j} + \sum_{i=3}^{n} d_{ij} c_{2i}^{(4)} = \quad\; 0 \cdots 0, \\ \cdots\cdots\cdots\cdots \\ d_{n-1,n} + d_{nn} c_{n-1,n}^{(2n-2)} = \qquad\quad 0. \end{cases}$$

3. The square-root or Cholesky's method (see [5; § 20])

This method can be used to solve (a) only when A is non-singular and symmetric. In this case, the following triangular decomposition for A is used:

$$A = S^T S,$$

where

$$S = \begin{bmatrix} s_{11} & s_{12} & \cdots & s_{1n} \\ 0 & s_{22} & \cdots & s_{2n} \\ \cdot & \cdot & \cdot & \cdot \\ 0 & 0 & \cdots & s_{nn} \end{bmatrix}.$$

Consequently, the s_{ij} can be determined using

$$s_{11} = \sqrt{a_{11}}, \quad s_{1j} = \frac{a_{1j}}{s_{11}},$$

$$s_{ii} = \sqrt{a_{ii} - \sum_{m=1}^{i-1} s_{mi}^2} \qquad (i > 1),$$

$$s_{ij} = \frac{a_{ij} - \sum_{m=1}^{i-1} s_{mi} s_{mj}}{s_{ii}} \qquad (j > i),$$

$$s_{ij} = 0 \qquad (i > j).$$

Once S has been determined,

$$S^T y = b,$$

and

$$S x = y$$

are solved by forward and back substitution, respectively.

4. The method of orthogonalisation

In this method, a series of transformations is used to reduce (a) to an orthogonal system

$$A^{(n)} x = b^{(n)},$$

in which the rows of $A^{(n)}$ satisfy the condition

$$\sum_{s=1}^{n} a_{is}^{(n)} a_{js}^{(n)} = 0 \qquad (i \neq j).$$

The transformations are defined inductively by

$$\lambda_{i\,m-1} = \frac{\sum_{k=1}^{n} a_{ik}^{(m-1)} a_{m-1\,k}^{(m-1)}}{\sum_{k=1}^{n} (a_{m-1\,k}^{(m-1)})^2},$$

$$a_{ij}^{(m)} = a_{ij}^{(m-1)} - \lambda_{im-1}\, a_{m-1\,j}^{(m-1)},$$

$$b_{i}^{(m)} = b_{i}^{(m-1)} - \lambda_{im-1}\, b_{m-1\,j}^{(m-1)},$$

where $m = 2, 3, \ldots, n$, $i = m, m+1, \ldots, n$, $j = 1, 2, \ldots, n$, and the unknowns $x_1, x_2, \ldots, x_n$ by

$$x_i = \sum_{t=1}^{n} \frac{a_{ti}^{(n)} b_t^{(n)}}{\sum_{s=1}^{n} (a_{ts}^{(n)})^2} \qquad (i = 1, 2, \ldots, n).$$

*Iterative Methods**)

5. A simple iterative method

Suppose that (a) is transformed to the form

$$\begin{cases} x_1 = c_{11} x_1 + c_{12} x_2 + \cdots + c_{1n} x_n + f_1, \\ x_2 = c_{21} x_1 + c_{22} x_2 + \cdots + c_{2n} x_n + f_2, \\ \cdots\cdots\cdots\cdots\cdots\cdots \\ x_n = c_{n1} x_1 + c_{n2} x_2 + \cdots + c_{nn} x_n + f_n, \end{cases}$$

which can be rewritten in matrix notation as

$$x = C x + f.$$

Starting with an arbitrary vector x_0, we define inductively the iterative process

$$x_m = C x_{m-1} + f \qquad (m = 1, 2, \ldots).$$

If $x_m \to x$ as $m \to \infty$, the vector x is the required solution of (a). A sufficient condition for the convergence of this iterative process is

$$\sum_{j=1}^{n} |c_{ij}| \leqq \mu < 1 \qquad (i = 1, 2, \ldots, n).$$

6. The Gauss-Seidel method

In the Gauss-Seidel method the approximations

$$x_1^{(k+1)}, x_2^{(k+1)}, \ldots, x_{i-1}^{(k+1)} \qquad \text{for} \qquad x_1, x_2, \ldots, x_{i-1}$$

at the $(k+1)$th stage of the iteration are used to calculate the value of x_i at that stage:

$$x_i^{(k+1)} = -\frac{1}{a_{ii}}\left[\sum_{j=1}^{i-1} a_{ij} x_j^{(k+1)} + \sum_{j=i+1}^{n} a_{ij} x_j^{(k)} - b_i\right].$$

*) see [6; chapter 9], [5; chapter 3] and [4; chapter 4].

Conditions for the convergence of this method are discussed in [5; § 38] and [6; § 9.7–2].

Methods for the inversion of matrices)*

7. The reinforcement method

Starting with A and the unit matrix I, the sequence of matrices

$$A^{(0)},\ B^{(1)},\ A^{(1)},\ B^{(2)},\ A^{(2)},\ \ldots,\ B^{(n)},\ A^{(n)},$$

is constructed inductively so that

$$A^{(0)} = A - I;$$

$$b_{ij}^{(m)} = \begin{cases} a_{ij}^{(m-1)}, & \text{if } i \neq m; \\ \delta_{ij}, & \text{if } i = m; \end{cases}$$

$$\delta_{ij} = \begin{cases} 1, & \text{if } i = j; \\ 0, & \text{if } i \neq j; \end{cases}$$

$$a_{ij}^{(m)} = b_{ij}^{(m)} - \frac{b_{im}^{(m)}\, a_{mj}^{(m-1)}}{1 + a_{mm}^{(m-1)}} \qquad (i, j, m = 1, 2, \ldots, n).$$

Consequently

$$A^{(n)} = A^{-1}.$$

8. An iterative method

The inverse A^{-1} can be determined using the iterative process

$$A_{k+1}^{-1} = A_k^{-1}\,(2\,I - A A_k^{-1}) \qquad (k = 0, 1, 2, \ldots), \tag{26.5}$$

where A_k^{-1} denotes the k-the approximation to A^{-1}. This method is used when an approximation, A_0^{-1}, to A^{-1} is known and a refinement of it is required. The method converges if A_0^{-1} is a sufficiently close approximation to A^{-1}.

9. Inversion by partitioning

In this method, the inverse of a matrix of high order is determined using the inverses of matrices of lower order.

Let A be partitioned so that

$$A = \left[\begin{array}{c|c} P & Q \\ \hline R & S \end{array}\right]$$

*) see [5; §§ 23–28] and [3; chapter 5].

where P and S are square matrices of order r and s $(r + s = n)$, respectively, and Q and R are rectangular matrices of order $r \times s$ and $s \times r$, respectively.
If A^{-1} is assumed to have the form

$$A^{-1} = \left[\begin{array}{c|c} \bar{P} & \bar{Q} \\ \hline \bar{R} & \bar{S} \end{array}\right],$$

where $\bar{P}$, $\bar{Q}$, $\bar{R}$ and $\bar{S}$ have the same size as P, Q, R and S, respectively, then

$$\begin{aligned} \bar{P} &= P^{-1} + XZ^{-1}Y \\ \bar{Q} &= -XZ^{-1} \\ \bar{R} &= -Z^{-1}Y \\ \bar{S} &= Z^{-1} \end{aligned}$$

where

$$X = P^{-1}Q, \quad Y = RP^{-1}, \quad Z = S - RX.$$

It is recommended that this method be used in conjunction with refinement methods (see method 8 above).

PROBLEMS

Compute the determinants of the following matrices:

1.
$$\begin{bmatrix} 2.37415 & 1.32457 & 2.25358 & 1.35877 & 0.58573 \\ 1.47132 & 2.24315 & 3.53837 & 2.58371 & 1.83921 \\ 4.17354 & 2.12839 & 1.78538 & 0.37859 & 3.24578 \\ 1.37531 & 0.28453 & 2.58389 & 3.57352 & 2.54783 \\ 1.57893 & 4.38921 & 2.55431 & 1.73894 & 3.47579 \end{bmatrix}.$$

2.
$$\begin{bmatrix} 0 & 2.1 & 3.2 & 1.2 & 0 \\ 0.8 & 0 & 2.9 & 0 & 1.4 \\ 1.4 & 2.5 & 0 & 3.2 & 2.8 \\ 1.6 & 0 & 1.3 & 0 & 1.1 \\ 0 & 1.8 & 3.8 & 2.2 & 0 \end{bmatrix}.$$

3.
$$\begin{bmatrix} 1.3259 & 2.3782 & 0 & 0 & 0 \\ 3.1112 & 4.0232 & 1.9982 & 0 & 0 \\ 0 & 5.0349 & 2.1122 & 6.3213 & 0 \\ 0 & 0 & 1.0932 & 4.9931 & 5.6632 \\ 0 & 0 & 0 & 6.0144 & 2.6420 \end{bmatrix}.$$

4. $$\begin{bmatrix} 1 & 2 & 4 & 8 & 16 \\ 1 & 0.4 & 0.16 & 0.064 & 0.0256 \\ 1 & -3 & 9 & -27 & 81 \\ 1 & 0.5 & 0.25 & 0.125 & 0.0625 \\ 1 & -0.6 & 0.36 & -0.216 & 0.1296 \end{bmatrix}.$$

5. $$\begin{bmatrix} 0.889 & 0.774 & -0.913 & 0.398 & 0.456 & 0.665 \\ 0.665 & 0.889 & 0.774 & -0.913 & 0.398 & 0.456 \\ 0.456 & 0.665 & 0.889 & 0.774 & -0.913 & 0.398 \\ 0.398 & 0.456 & 0.665 & 0.889 & 0.774 & -0.913 \\ -0.913 & 0.398 & 0.456 & 0.665 & 0.889 & 0.774 \\ 0.774 & -0.913 & 0.398 & 0.456 & 0.665 & 0.889 \end{bmatrix}.$$

6. $$\begin{bmatrix} 1.324 & -0.998 & 1.112 & 0.456 & 0.695 & 1.325 \\ 0.998 & 0.345 & 1.425 & 0.546 & 1.965 & 1.782 \\ -1.112 & -1.425 & 0.832 & 1.387 & -0.695 & 0.611 \\ -0.456 & -0.546 & -1.387 & 1.327 & 0.874 & 1.883 \\ -0.695 & -1.965 & 0.695 & -0.874 & 0.832 & 1.913 \\ -1.325 & -1.782 & -0.611 & -1.883 & -1.913 & -1.139 \end{bmatrix}.$$

7. $$\begin{bmatrix} 1.2 & 1.4 & -1.9 & 1.5 & -0.9 & 2.1 \\ 0.8 & 1.3 & 2.6 & -0.9 & 1.5 & 2.3 \\ 0 & 2.4 & 1.5 & 0.6 & 2.3 & 0.7 \\ 0 & 0 & 1.1 & 2.1 & 0.8 & 1.5 \\ 0 & 0 & 0 & 1.4 & 0.8 & 2.5 \\ 0 & 0 & 0 & 0 & 1.7 & 2.1 \end{bmatrix}.$$

8. $$\begin{bmatrix} 1.8724 & 1.2451 & 0 & 0 & 0 & 0 \\ 0.9812 & -0.7812 & 0 & 0 & 0 & 0 \\ 0 & 0 & 0.5426 & 0.6324 & 0 & 0 \\ 0 & 0 & 0.4326 & -0.7825 & 0 & 0 \\ 0 & 0 & 0 & 0 & 0.5546 & -0.6254 \\ 0 & 0 & 0 & 0 & 0.3824 & 0.4562 \end{bmatrix}.$$

9. $$\begin{bmatrix} 0.9345 & 0.5652 & 1.6253 & 0.7341 & 1.8752 & -0.9829 & 0.4565 \\ 0.5652 & 1.4254 & 0.8059 & 1.4091 & -0.5312 & -1.1391 & 1.3445 \\ 1.6253 & 0.8059 & 1.3492 & 1.4356 & 0.6274 & 0.7709 & 1.3214 \\ 0.7341 & 1.4091 & 1.4356 & 0.9661 & -1.7342 & 1.3208 & 0.7613 \\ 1.8752 & -0.5312 & 0.6274 & -1.7342 & 0.7915 & 0.8814 & 1.4707 \\ -0.9829 & -1.1391 & 0.7709 & 1.3208 & 0.8814 & 0.8059 & -1.6253 \\ 0.4565 & 1.3445 & 1.3212 & 0.7613 & 1.4707 & -1.6253 & 0.7915 \end{bmatrix}.$$

10.
$$\begin{bmatrix} 0.44 & 1.31 & 0 & 0 & 0 & 0 & 0 \\ 1.35 & -0.86 & 0 & 0 & 0 & 0 & 0 \\ 0 & 0 & 0.34 & 0.21 & 1.13 & 0 & 0 \\ 0 & 0 & 0.72 & -0.98 & 0.75 & 0 & 0 \\ 0 & 0 & 0.83 & 1.27 & -0.54 & 0 & 0 \\ 0 & 0 & 0 & 0 & 0 & 0.45 & 0.18 \\ 0 & 0 & 0 & 0 & 0 & 0.31 & -0.16 \end{bmatrix}.$$

11.
$$\begin{bmatrix} 2.15 & 3.18 & 9.18 & 3.15 & 6.82 & 4.21 & 7.62 \\ 5.64 & 3.24 & 8.21 & 7.61 & 8.32 & 6.15 & 6.24 \\ 8.78 & 9.13 & 5.18 & 6.18 & 7.29 & 8.63 & 3.21 \\ 8.62 & 3.63 & 8.56 & 4.28 & 8.42 & 8.63 & 5.21 \\ 3.81 & 4.24 & 3.21 & 8.56 & 7.26 & 4.32 & 6.34 \\ 8.46 & 5.68 & 7.86 & 7.51 & 3.18 & 7.15 & 9.36 \\ 5.83 & 6.37 & 8.94 & 6.18 & 7.56 & 9.73 & 2.31 \end{bmatrix}.$$

12.
$$\begin{bmatrix} 0 & -1.8 & 2.3 & 1.4 & 3.2 & -1.5 & 0 \\ 1.4 & 0.8 & 2.6 & -1.7 & 1.1 & 0.4 & 2.1 \\ 3.1 & 1.7 & 0 & 1.3 & 0 & -2.8 & 2.3 \\ 2.5 & -3.3 & 1.4 & 0.3 & 1.6 & 2.1 & 0.8 \\ 2.3 & -4.1 & 0 & 1.5 & 0 & -3.2 & 1.2 \\ 1.5 & 0.7 & 4.1 & 3.3 & 3.2 & 0.9 & 2.1 \\ 0 & 1.4 & -1.9 & 2.7 & 1.8 & -3.1 & 0 \end{bmatrix}.$$

13.
$$\begin{bmatrix} 0.42 & 1.34 & 0.88 & 0.75 & 1.21 & 0.39 & -2.11 & 1.52 \\ 1.34 & 0.35 & 1.87 & 0.43 & 0.54 & 1.25 & 0.75 & 0.91 \\ 0.88 & 1.87 & 0.38 & 0.46 & 1.71 & 0.41 & 1.87 & 0.19 \\ 0.75 & 0.43 & 0.46 & 1.84 & -2.14 & 0.36 & 1.13 & -2.54 \\ 1.21 & 0.54 & 1.71 & -2.14 & 0.89 & 1.71 & 0.39 & 0.87 \\ 0.39 & 1.25 & 0.41 & 0.36 & 1.71 & -2.11 & 0.42 & 1.14 \\ -2.11 & 0.75 & 1.87 & 1.13 & 0.39 & 0.42 & 1.31 & 0.45 \\ 1.52 & 0.91 & 0.19 & -2.54 & 0.87 & 1.14 & 0.45 & 1.75 \end{bmatrix}.$$

14.
$$\begin{bmatrix} 1 & 2 & 0.61 & 0 & 0 & 0 & 0 & 0 \\ 0.32 & 1.67 & 0.72 & 0 & 0 & 0 & 0 & 0 \\ 1.56 & 1.24 & 0.32 & 0 & 0 & 0 & 0 & 0 \\ 0 & 0 & 0 & 1.32 & -2.43 & 0 & 0 & 0 \\ 0 & 0 & 0 & 1.25 & 1.62 & 0 & 0 & 0 \\ 0 & 0 & 0 & 0 & 0 & 0.19 & 1.71 & 2.13 \\ 0 & 0 & 0 & 0 & 0 & 1.32 & 2.61 & 0.98 \\ 0 & 0 & 0 & 0 & 0 & 2.52 & 0.81 & 1.21 \end{bmatrix}.$$

15. $$\begin{bmatrix} 1.21 & 0 & 0 & 0 & 0 & 0 & 0 & 0.32 \\ 0 & -0.85 & 1.27 & 0.77 & 1.32 & 0.84 & 1.45 & 0 \\ 0 & 0.67 & 1.41 & 0.76 & 0.88 & 1.23 & 0.11 & 0 \\ 0 & 0.24 & 1.12 & -2.43 & 1.99 & 0.91 & 0.83 & 0 \\ 0 & 1.24 & 0.32 & 1.54 & 0.32 & 1.39 & 1.51 & 0 \\ 0 & 1.25 & 0.62 & 1.77 & 0.19 & 1.18 & 0.71 & 0 \\ 0 & 1.56 & 0.67 & 2.13 & 1.52 & 0.32 & 1.31 & 0 \\ 1.52 & 0 & 0 & 0 & 0 & 0 & 0 & 0.81 \end{bmatrix}.$$

16. $$\begin{bmatrix} 0.3 & 1.1 & -0.8 & 0.4 & 1.5 & 2.1 & 0.7 & -2.5 \\ -2.5 & 0.3 & 1.1 & -0.8 & 0.4 & 1.5 & 2.1 & 0.7 \\ 0.7 & -2.5 & 0.3 & 1.1 & -0.8 & 0.4 & 1.5 & 2.1 \\ 2.1 & 0.7 & -2.5 & 0.3 & 1.1 & -0.8 & 0.4 & 1.5 \\ 1.5 & 2.1 & 0.7 & -2.5 & 0.3 & 1.1 & -0.8 & 0.4 \\ 0.4 & 1.5 & 2.1 & 0.7 & -2.5 & 0.3 & 1.1 & -0.8 \\ -0.8 & 0.4 & 1.5 & 2.1 & 0.7 & -2.5 & 0.3 & 1.1 \\ 1.1 & -0.8 & 0.4 & 1.5 & 2.1 & 0.7 & -2.5 & 0.3 \end{bmatrix}.$$

17. $$\begin{bmatrix} 1.1 & 1.2 & -2.3 & 0.8 & 1.5 & 0.6 & 1.7 & 0.7 & -2.3 \\ -1.2 & 1.7 & 0.4 & 1.8 & 2.2 & 0.5 & -1.2 & 0.3 & 1.7 \\ 2.3 & -0.4 & 2.5 & 0.9 & 1.4 & 2.6 & 1.7 & 0.6 & 0.8 \\ -0.8 & -1.8 & -0.9 & -2.4 & 1.2 & -1.4 & -2.9 & -1.4 & 1.3 \\ -1.5 & -2.2 & -1.4 & -1.2 & 0.3 & 1.9 & 2.1 & 0.2 & -2.3 \\ -0.6 & -0.5 & -2.6 & 1.4 & -1.9 & 0.8 & 1.4 & 0.5 & -2.5 \\ -1.7 & 1.2 & -1.7 & 2.9 & -2.1 & -1.4 & 0.7 & 1.6 & 1.1 \\ -0.7 & -0.3 & -0.6 & 1.4 & -0.2 & -0.5 & -1.6 & 1.3 & 2.8 \\ 2.3 & -1.7 & -0.8 & -1.3 & 2.3 & 2.5 & -1.1 & -2.8 & -1.9 \end{bmatrix}.$$

18. $$\begin{bmatrix} 1.21 & 0.31 & 1.23 & 2.32 & 0.84 & -1.99 & 2.12 & 1.77 & -1.91 \\ 0.61 & 0.97 & 1.45 & 0.65 & -2.23 & 0.35 & 1.23 & 0.54 & 0.33 \\ 0 & -0.41 & 1.14 & 2.24 & 0.67 & 1.48 & -2.11 & 0.78 & 1.87 \\ 0 & 0 & 1.43 & 0.61 & -1.36 & 0.72 & 1.48 & 1.15 & 0.56 \\ 0 & 0 & 0 & -1.35 & 0.44 & 1.16 & -2.61 & 0.98 & 1.91 \\ 0 & 0 & 0 & 0 & 0.34 & 1.56 & 1.21 & 0.95 & 0.87 \\ 0 & 0 & 0 & 0 & 0 & 0.92 & 1.32 & 2.14 & 0.74 \\ 0 & 0 & 0 & 0 & 0 & 0 & 0.29 & -1.65 & 2.55 \\ 0 & 0 & 0 & 0 & 0 & 0 & 0 & 0.99 & -1.89 \end{bmatrix}.$$

19.
$$\begin{bmatrix}
1.1 & 1 & 0 & 0 & 0 & 0 & 0 & 0 & 0 \\
-1 & 0.9 & 1 & 0 & 0 & 0 & 0 & 0 & 0 \\
0 & -1 & 1.2 & 1 & 0 & 0 & 0 & 0 & 0 \\
0 & 0 & -1 & 0.8 & 1 & 0 & 0 & 0 & 0 \\
0 & 0 & 0 & -1 & 0.7 & 1 & 0 & 0 & 0 \\
0 & 0 & 0 & 0 & -1 & 2.3 & 1 & 0 & 0 \\
0 & 0 & 0 & 0 & 0 & -1 & 1.9 & 1 & 0 \\
0 & 0 & 0 & 0 & 0 & 0 & -1 & 0.5 & 1 \\
0 & 0 & 0 & 0 & 0 & 0 & 0 & -1 & 2.1
\end{bmatrix}.$$

20.
$$\begin{bmatrix}
0.625 & 0.343 & 0.835 & 0.841 & -0.491 & 0.181 & 0.371 & 0.972 & -0.661 \\
0.343 & 1.731 & 1.322 & 1.149 & 1.213 & 1.212 & 1.341 & -1.181 & 1.122 \\
0.835 & 1.322 & -0.665 & 0.535 & -0.543 & 0.395 & 0.723 & 0.121 & 0.167 \\
0.841 & 1.149 & 0.535 & -1.214 & 1.982 & 1.131 & 1.171 & -1.348 & -1.142 \\
-0.491 & 1.213 & -0.543 & 1.982 & 0.381 & 0.137 & 0.234 & 0.195 & -0.625 \\
0.181 & 1.212 & 0.395 & 1.131 & 0.137 & 1.551 & 1.621 & 1.124 & 1.137 \\
0.371 & 1.341 & 0.723 & 1.171 & 0.234 & 1.621 & 0.841 & 0.627 & 0.222 \\
0.972 & -1.181 & 0.121 & -1.348 & 0.195 & 1.124 & 0.627 & -0.585 & 1.721 \\
-0.661 & 1.122 & 0.167 & -1.142 & -0.625 & 1.137 & 0.222 & 1.721 & 0.149
\end{bmatrix}.$$

21.
$$\begin{bmatrix}
0 & 0.1 & 0.2 & 0.5 & 1.2 & 0.8 & 1.8 & 1.3 & 1 & 0 \\
0.9 & 0.6 & 0.1 & 0.7 & -2.2 & 0.4 & 0.3 & 1.2 & 1.1 & 0.5 \\
1.3 & 0.4 & 0 & 1.4 & 0.5 & 1.3 & 2.1 & 0 & 1.4 & 0.7 \\
1.2 & -2.7 & 1.9 & 0.4 & 0.6 & 1.7 & 0.1 & 0.2 & 0.8 & 1.3 \\
0.5 & 1.4 & 0.6 & 1.2 & 0 & 0 & 0.9 & 1.4 & 0.1 & -2.3 \\
2.3 & 0.8 & 1.9 & 2.1 & 0 & 0 & 0.7 & 1.7 & 1.2 & 1.5 \\
0.5 & 0.7 & 1.2 & 1.1 & 1.8 & -2.9 & 2 & 0.2 & 1.5 & 1.6 \\
2.4 & 1.4 & 0 & 2.1 & 0.7 & 0.9 & 0.1 & 0 & 1.5 & 0.2 \\
1.6 & 0.2 & 2.3 & 1.9 & 2.3 & 0.6 & 2 & 1.1 & 1.8 & 1.6 \\
0 & 1.8 & 0.9 & 1.9 & 2.1 & 0.5 & 1.6 & 2.2 & 1.6 & 0
\end{bmatrix}.$$

22.
$$\begin{bmatrix}
1.3 & 3.4 & 0.5 & 1.2 & 1.1 & 1.4 & 0.3 & 0.6 & 1.2 & -1.8 \\
0.8 & 1.1 & 0.7 & 2.9 & 1.8 & -2.7 & 0.3 & 1.9 & 2.4 & 0.8 \\
0 & 0.3 & 1.6 & 0.7 & -2.3 & 1.1 & 2.2 & 0.6 & 1.3 & 0.7 \\
0 & 0 & -2.1 & 0.8 & 0.9 & 1.5 & 1.6 & 2.3 & 1.9 & 3.6 \\
0 & 0 & 0 & -0.8 & 1.7 & 2.8 & 1.9 & 0.5 & 1.3 & -0.5 \\
0 & 0 & 0 & 0 & 0.9 & 1.8 & 1.5 & 2.8 & 1.7 & 2.1 \\
0 & 0 & 0 & 0 & 0 & 1.9 & 0.2 & 1.3 & 0.8 & -1.9 \\
0 & 0 & 0 & 0 & 0 & 0 & 1.4 & 2.5 & 0.6 & 2.1 \\
0 & 0 & 0 & 0 & 0 & 0 & 0 & 0.9 & 2.3 & 1.6 \\
0 & 0 & 0 & 0 & 0 & 0 & 0 & 0 & 1.5 & 0.8
\end{bmatrix}.$$

23.
$$\begin{bmatrix} 1.21 & 3.21 & 0.98 & 0 & 0 & 0 & 0 & 0.33 & -1.45 & 0.87 \\ 2.31 & 2.01 & -0.44 & 0 & 0 & 0 & 0 & 1.46 & 2.87 & -1.56 \\ 1.43 & 0.62 & 0.55 & 0 & 0 & 0 & 0 & 0.32 & 1.67 & 1.45 \\ 0 & 0 & 0 & 3.01 & 2.24 & -0.39 & 1.44 & 0 & 0 & 0 \\ 0 & 0 & 0 & 2.11 & 1.46 & -0.66 & 2.14 & 0 & 0 & 0 \\ 0 & 0 & 0 & 0.62 & -1.25 & 0.59 & 1.81 & 0 & 0 & 0 \\ 0 & 0 & 0 & 1.84 & 2.39 & 3.84 & 2.05 & 0 & 0 & 0 \\ 2.32 & 1.65 & 0.75 & 0 & 0 & 0 & 0 & 1.32 & 2.31 & 0.54 \\ -2.62 & 0.46 & 1.16 & 0 & 0 & 0 & 0 & 0.81 & 1.84 & -2.43 \\ 0.69 & 1.95 & -2.18 & 0 & 0 & 0 & 0 & 3.18 & 0.84 & 1.54 \end{bmatrix}.$$

24.
$$\begin{bmatrix} 0.334 & 0.321 & 0.148 & -0.172 & 0.232 & 0.731 & -0.177 & 0.123 & 0.840 & -0.173 \\ -0.321 & 1.249 & 1.321 & -1.214 & 1.247 & -1.991 & 1.842 & 1.148 & -1.341 & 1.425 \\ -0.148 & -1.321 & 0.155 & 0.211 & 0.259 & -0.778 & -0.992 & 0.256 & 0.625 & 0.431 \\ 0.172 & 1.214 & -0.211 & 1.952 & 1.256 & 1.841 & 1.531 & 1.177 & 1.232 & 1.625 \\ -0.232 & -1.247 & -0.259 & -1.256 & 0.482 & 0.911 & 0.771 & 0.841 & 0.991 & -0.482 \\ -0.731 & 1.991 & 0.778 & -1.841 & -0.911 & 1.321 & 1.155 & 1.211 & 1.256 & 1.212 \\ 0.177 & -1.842 & 0.992 & -1.531 & -0.771 & -1.155 & 0.232 & 0.781 & 0.173 & -0.841 \\ -0.123 & -1.148 & -0.256 & -1.177 & -0.841 & -1.211 & -0.781 & 1.187 & -1.958 & -1.909 \\ -0.840 & 1.341 & -0.625 & -1.232 & -0.991 & -1.256 & -0.173 & 1.958 & 0.324 & 1.211 \\ 0.173 & -1.425 & -0.431 & -1.625 & 0.482 & -1.212 & 0.841 & 1.909 & -1.211 & 0.448 \end{bmatrix}$$

Using methods **9** and **7**, invert the following matrices:

25.
$$\begin{bmatrix} 1 & 1 & 0 & 0 \\ -1 & 1 & 1 & 0 \\ 0 & -1 & 1 & 1 \\ 0 & 0 & -1 & 1 \end{bmatrix}.$$

26.
$$\begin{bmatrix} 1.31 & -2.22 & 1.56 & -2.31 & 1.33 \\ -2.45 & 1.23 & -2.91 & 1.98 & 1.78 \\ 0 & 1.77 & -2.79 & 1.76 & -2.65 \\ 0 & 0 & 1.32 & -2.33 & 1.11 \\ 0 & 0 & 0 & 1.31 & -2.29 \end{bmatrix}.$$

27.
$$\begin{bmatrix} 3.1 & 0.2 & 4 & -7.1 & -8.4 & 2.2 \\ -7.4 & -6.3 & 3.6 & 1.4 & 0.9 & 3.4 \\ 3.6 & 0.4 & 1.7 & 4.8 & 3.3 & -8.7 \\ 6.4 & 3.7 & 3.1 & 5.6 & 3.7 & -6.1 \\ -7.8 & 5.6 & 2.9 & 6.7 & 4 & 1.9 \\ -5 & 0.8 & 1 & 3.5 & 2.9 & 2 \end{bmatrix}.$$

28.
$$\begin{bmatrix} 5.39 & 7.86 & 5.63 & 3.21 & 9.87 & 5.78 & 3.21 \\ 1.19 & 8.56 & 7.38 & 9.31 & 9.82 & 3.21 & 8.67 \\ 5.67 & 8.29 & 9.32 & 5.45 & 8.48 & 9.65 & 9.67 \\ 9.53 & 7.53 & 8.64 & 9.31 & 9.87 & 5.64 & 3.87 \\ 9.32 & 9.16 & 8.29 & 9.32 & 7.56 & 3.24 & 9.28 \\ 7.64 & 5.68 & 6.78 & 9.31 & 3.84 & 8.79 & 6.27 \\ 6.54 & 3.29 & 4.68 & 9.76 & 4.29 & 6.57 & 6.84 \end{bmatrix}.$$

29.
$$\begin{bmatrix} 0.843 & 0.522 & 0.451 & 0.842 & -0.971 & 0.875 & 0 & 0.175 \\ 0.245 & 0.878 & 0 & 0.356 & 0.625 & -0.910 & 0.262 & 0.217 \\ 0 & 0.394 & 1.382 & -0.911 & 0 & 0.429 & 0.358 & 0.232 \\ 0.871 & 0.377 & 0.423 & 1.207 & 0.625 & 0.563 & 0 & 0.671 \\ 0.378 & 0.921 & 0.417 & 0.832 & 0.822 & 0.444 & 0.812 & 0.542 \\ 0.629 & 0.458 & 0 & 0.732 & -0.982 & 0.754 & 0.587 & 0.391 \\ 0 & 0.494 & 0.358 & 0.675 & 0.415 & 0.604 & 1.304 & -0.908 \\ 0.422 & 0.155 & 0.532 & 0 & 0.771 & 0.475 & 0.415 & 0.608 \end{bmatrix};$$

30.
$$\begin{bmatrix} 4 & 3 & 1 & 2 & 1.5 & 3 & -1.6 & 2 & 0.7 \\ 0.7 & 4 & 3 & 1 & 2 & 1.5 & 3 & -1.6 & 2 \\ 2 & 0.7 & 4 & 3 & 1 & 2 & 1.5 & 3 & -1.6 \\ -1.6 & 2 & 0.7 & 4 & 3 & 1 & 2 & 1.5 & 3 \\ 3 & -1.6 & 2 & 0.7 & 4 & 3 & 1 & 2 & 1.5 \\ 1.5 & 3 & -1.6 & 2 & 0.7 & 4 & 3 & 1 & 2 \\ 2 & 1.5 & 3 & -1.6 & 2 & 0.7 & 4 & 3 & 1 \\ 1 & 2 & 1.5 & 3 & -1.6 & 2 & 0.7 & 4 & 3 \\ 3 & 1 & 2 & 1.5 & 3 & -1.6 & 2 & 0.7 & 4 \end{bmatrix}.$$

For the following matrices, use method 7 to determine a first approximation A_0^{-1} {with elements rounded to two significant places}, which is refined using 8 so that all the elements of the inverse are accurate to five significant places.

31.
$$\begin{bmatrix} 0.3 & 0.8 & 0.7 & -0.9 \\ -0.9 & 0.3 & 0.8 & 0.7 \\ 0.7 & -0.9 & 0.3 & 0.8 \\ 0.8 & 0.7 & -0.9 & 0.3 \end{bmatrix}.$$

32.
$$\begin{bmatrix} 7 & 2 & -8 & 3 & 5 \\ -2 & 1 & 6 & -4 & 1 \\ 8 & -6 & 2 & 5 & -6 \\ -3 & 4 & -5 & 1 & 2 \\ -5 & -1 & 6 & -2 & 3 \end{bmatrix}.$$

33.
$$\begin{bmatrix} 3.325 & 2.988 & 1.212 & 0.556 & 0.675 & 1.425 \\ 0.988 & 2.455 & 3.325 & 0.536 & 1.865 & 1.789 \\ 1.312 & 1.435 & 2.852 & 3.487 & 0.675 & 0.631 \\ 0.556 & 0.446 & 1.381 & 3.325 & 2.864 & 1.863 \\ 0.795 & 1.865 & 0.795 & 0.854 & 2.842 & 3.923 \\ 1.425 & 1.712 & 0.612 & 1.884 & 1.923 & 3.339 \end{bmatrix}.$$

34.
$$\begin{bmatrix} 0.23 & 0.84 & 0.56 & 0.87 & 0.12 & 0.18 & 0.34 \\ 0.87 & 0.93 & 0.97 & 0.87 & 0.87 & 0.56 & 0.76 \\ 0.31 & 0.23 & 0.45 & 0.64 & 0.87 & 0.56 & 0.64 \\ 0.32 & 0.38 & 0.28 & 0.51 & 0.23 & 0.31 & 0.38 \\ 0.56 & 0.78 & 0.76 & 0.71 & 0.87 & 0.56 & 0.76 \\ 0.87 & 0.59 & 0.39 & 0.29 & 0.45 & 0.43 & 0.48 \\ 0.44 & 0.34 & 0.38 & 0.83 & 0.89 & 0.93 & 0.29 \end{bmatrix}.$$

35.
$$\begin{bmatrix} 1.5 & 2 & 4 & 8 & 16 & 32 & 64 & 128 \\ 1 & 1.8 & 1.69 & 2.20 & 2.86 & 3.71 & 4.83 & 6.27 \\ 1 & 2.1 & 4.91 & 9.26 & 19.45 & 40.84 & 85.77 & 180.11 \\ 1 & 1.5 & 2.25 & 3.88 & 5.06 & 7.59 & 11.39 & 17.09 \\ 1 & 7.7 & 2.89 & 4.91 & 8.85 & 14.2 & 24.14 & 41.03 \\ 1 & 0.9 & 0.81 & 0.73 & 0.66 & 1.09 & 0.53 & 0.48 \\ 1 & 0.8 & 0.64 & 0.51 & 0.41 & 0.33 & 0.76 & 0.21 \\ 1 & 1.6 & 2.56 & 4.10 & 6.55 & 10.49 & 16.78 & 27.34 \end{bmatrix}.$$

36.
$$\begin{bmatrix} 1.91 & 0 & 0.32 & 0 & 1.42 & 0.44 & 0.64 & 1.19 & 0 \\ 1.71 & 1.32 & -1.21 & 0.99 & 1.11 & -0.49 & -1.18 & 0.23 & 0.65 \\ 0.48 & 1.21 & 1 & 0.52 & 0.29 & -1.44 & 0 & 1.26 & 0.81 \\ -1.39 & 0.53 & 0.85 & 2.24 & 0.34 & -1.15 & 0.84 & 1.17 & 0.31 \\ 0 & 0.43 & 0.21 & 1.25 & 1.36 & 1.12 & 0.91 & 1.16 & 0.52 \\ -1.13 & 1.42 & 0.52 & -1.45 & 0 & 1.51 & 0.49 & 1.31 & 0.32 \\ 1.33 & 0.62 & -1.44 & 0.24 & 0.65 & 0.36 & 1 & -1.19 & 1.16 \\ 0.70 & 0 & 0.77 & 0 & 0.81 & 1.31 & 0.29 & 2.12 & 0.43 \\ 0.45 & 0 & 1.23 & 1.57 & 0.98 & 0 & 1.12 & 0.75 & 1.55 \end{bmatrix}.$$

37.
$$\begin{bmatrix} 2.3 & -2.3 & 0.5 & 0 & 0 & 0.4 & 0.6 & 0 & 1.7 & 0 \\ 0.9 & -0.8 & 1.5 & 0 & 0.6 & 0.9 & 0 & 0 & 0.4 & 0 \\ 0 & -2.1 & 1.7 & 1.4 & 0.3 & 0 & 0.1 & 0 & 0.8 & 0.3 \\ 2 & 1 & 1.2 & 1.4 & -2.2 & 0 & 0.7 & 0.2 & 0 & 0.1 \\ 0 & 0.8 & 0 & -2.8 & 1.7 & 1.6 & 0.5 & 0 & 0.3 & 0 \\ 0.6 & 0.4 & 1 & 0 & 1.5 & 1.9 & -2.5 & 0.6 & 0 & 0.8 \\ 0 & 2 & 0 & 0.2 & 0 & -2.1 & 1.3 & -2.3 & 0 & 0 \\ 0.5 & 0 & 0.7 & 0.1 & 0.9 & 0.4 & 0.6 & 2.7 & 1.4 & 1 \\ 0 & 0 & 2 & 0 & 1 & 0.1 & 0 & -2.6 & 1.3 & 1.5 \\ 0.7 & 0.9 & 0 & 1 & 0 & 0.8 & 2 & 1 & -2.1 & 1. \end{bmatrix}.$$

In examples 38–69, use the method suggested to solve either the algebraic system $Ax = b$ or $(C + kD)\,x = Ax = b$ for the values of k specified. For convenience, b is often written as the row vector b^T.

Solve the following systems using method **1** and compute the determinants of A $\{(C + kD)\}$.

38. $$D = \begin{bmatrix} 2.1 & -4.5 & -2.0 \\ 3.0 & 2.5 & 4.3 \\ -6.0 & 3.5 & 2.5 \end{bmatrix},\ C = \begin{bmatrix} 0.1 & 0 & 0 \\ 0 & 0.1 & 0 \\ 0 & 0 & 0.1 \end{bmatrix},$$

$b^T = [19.07,\ 3.21,\ -18.25]$, $k = 0(1)15$.

39. $$D = \begin{bmatrix} 3 & 1 & 2 & 4 \\ -5 & 2 & 6 & 7 \\ 0 & 3 & 4 & 1 \\ 0 & 0 & -5 & -6 \end{bmatrix},$$

$$C = \begin{bmatrix} -0.9 & 0.7 & 0.8 & -0.5 \\ 0.6 & -0.5 & 0.4 & 0.3 \\ 0 & 0.3 & 0.8 & -0.9 \\ 0 & 0 & 0.7 & 0.6 \end{bmatrix},\ b = \begin{bmatrix} 4 \\ -13 \\ 1 \\ -5 \end{bmatrix},$$

$k = 0(1)15$.

40. $$D = \begin{bmatrix} 2.33 & 0.81 & 0.67 & 0.92 & -0.53 \\ -0.53 & 2.33 & 0.81 & 0.67 & 0.92 \\ 0.92 & -0.53 & 2.33 & 0.81 & 0.67 \\ 0.67 & 0.92 & -0.53 & 2.33 & 0.81 \\ 0.81 & 0.67 & 0.92 & -0.53 & 2.33 \end{bmatrix},$$

$$C = \begin{bmatrix} 0.2 & 0 & 0.2 & 0 & 0 \\ 0 & 0.2 & 0 & 0.2 & 0 \\ 0.2 & 0 & 0.2 & 0 & 0.2 \\ 0 & 0.2 & 0 & 0.2 & 0 \\ 0 & 0 & 0.2 & 0 & 0.2 \end{bmatrix},\ b = \begin{bmatrix} 4.2 \\ 4.2 \\ 4.2 \\ 4.2 \\ 4.2 \end{bmatrix},$$

$k = 0(1)15$.

41. $$D = \begin{bmatrix} 0.87 & 0.34 & 0.44 & 0.56 & 0.65 & 0.37 \\ 1.39 & 0.77 & 1.28 & 0.82 & 0.49 & 0 \\ 0.44 & 1.28 & 0 & 0.85 & 0.78 & 1.33 \\ 0 & 0.82 & 0.85 & 1.27 & 0.34 & 0.72 \\ 0.65 & 0 & 0.78 & 1.34 & 0.55 & 0 \\ 1.37 & 0.99 & 0.33 & 0.72 & 0 & 0.43 \end{bmatrix},$$

$$C = \begin{bmatrix} 0.3 & 0 & 0 & 0 & 0 & 0 \\ 0 & 0.3 & 0 & 0 & 0 & 0 \\ 0 & 0 & 0.3 & 0 & 0 & 0 \\ 0 & 0 & 0 & 0.3 & 0 & 0 \\ 0 & 0 & 0 & 0 & 0.3 & 0 \\ 0 & 0 & 0 & 0 & 0 & 0.3 \end{bmatrix}, \qquad b = \begin{bmatrix} 1.08 \\ 0.60 \\ 1.22 \\ -0.51 \\ 0.42 \\ 1.04 \end{bmatrix},$$

$k = 0(0.2)3.0.$

42. $$A = \begin{bmatrix} 0.411 & 0.421 & -0.333 & 0.313 & -0.141 & -0.381 & 0.245 \\ 0.421 & 0.705 & 0.139 & -0.409 & 0.321 & 0.625 & 0.101 \\ 0.123 & -0.239 & 0.502 & 0.901 & 0.243 & 0.819 & 0.321 \\ 0.413 & 0.309 & 0.801 & 0.865 & 0.423 & 0.118 & 0.183 \\ 0.241 & -0.221 & -0.243 & 0.134 & 1.274 & 0.712 & 0.423 \\ 0.281 & 0.525 & 0.719 & 0.118 & -0.974 & 0.808 & 0.923 \\ 0.246 & -0.301 & 0.231 & 0.813 & -0.702 & 1.223 & 1.105 \end{bmatrix},$$

$b^T = [0.096, \quad 1.252, \quad 1.024, \quad 1.023, \quad 1.155, \quad 1.937, \quad 1.673].$

43. $$A = \begin{bmatrix} 1.21 & 1.43 & 1.32 & 0 & 0.84 & 0.24 & 1.12 & 0.82 \\ 0 & 2.14 & -0.85 & 1.27 & 0.77 & 1.32 & 1.34 & 1.45 \\ 1.23 & 0.67 & 1.41 & 0.76 & 0.14 & 0.5 & 0.78 & 1.23 \\ 0.11 & 0.24 & 1.12 & 0 & 2.49 & 0.91 & 0.83 & 1.43 \\ 1.32 & 1.24 & 0.32 & 2.04 & 0.32 & 1.39 & 1.51 & 0 \\ 1.24 & 1.15 & 1.12 & 0 & 0.19 & 1.18 & 0.71 & 1.77 \\ 0 & 2.06 & 0.67 & 2.13 & 1.52 & 0.32 & 1.31 & 1.42 \\ 2.02 & 1.14 & 0 & 1.15 & 1.23 & 2.14 & 0 & 0.81 \end{bmatrix},$$

$b^T = [6.48, \quad 6.94, \quad 6.22, \quad 6.63, \quad 7.64, \quad 6.86, \quad 8.93, \quad 7.99].$

44. $$A = \begin{bmatrix} 1.61 & 0.31 & 1.23 & 2.32 & 0.84 & -1.99 & 2.12 & 1.77 & -1.91 \\ 0.61 & 1.37 & 1.45 & 0.65 & -2.23 & 0.35 & 1.23 & 0.54 & 0.33 \\ 0 & -0.41 & 1.14 & 2.64 & 0.67 & 1.48 & -2.11 & 0.78 & 1.87 \\ 0 & 0 & 1.43 & 0.61 & -0.96 & 0.72 & 1.48 & 1.15 & 0.56 \\ 0 & 0 & 0 & -1.35 & 0.44 & 1.50 & 1.16 & -2.61 & 0.98 \\ 1.91 & 0 & 0 & 0 & 0 & 0.34 & 1.96 & 1.21 & 0.95 \\ 0.80 & 0.87 & 0 & 0 & 0 & 0 & 0 & 1.32 & 1.32 \\ 2.14 & 0.74 & 0 & 0 & 0 & 0 & 0 & 0 & 0.69 \\ -1.65 & 2.55 & 0 & 0 & 0 & 0 & 0 & 0 & 0 \end{bmatrix},$$

$b^T = [5.9, \quad 3.9, \quad 5.66, \quad 4.59, \quad 0.53, \quad 4.93, \quad 5.12, \quad 1.19, \quad -0.09].$

45. $$A = \begin{bmatrix} 1.3 & -2.3 & 0.5 & 0 & 0 & 0.4 & 0.6 & 0 & 1.7 & 0 \\ 0.9 & -1.8 & 1.5 & 0 & 0.6 & 0.9 & 0 & 0 & 0.4 & 0 \\ 0 & -2.1 & 0.7 & 1.4 & 0.3 & 0 & 0.1 & 0 & 0.8 & 0.3 \\ 2 & 1 & 1.2 & 0.4 & -2.2 & 0 & 0.7 & 0.2 & 0 & 0.1 \\ 0 & 0.8 & 0 & -2.8 & 0.7 & 1.6 & 0.5 & 0 & 0.3 & 0 \\ 0.6 & 0.4 & 1 & 0 & 1.5 & 0.9 & 0.9 & -2.5 & 0.6 & 0 \\ 0.8 & 0 & 2 & 0 & 0.2 & 0 & -2.1 & 0.3 & -2.3 & 0 \\ 0 & 0.5 & 0 & 0.7 & 0.1 & 0.9 & 0.4 & 0.6 & 1.7 & 1.4 \\ 1 & 0 & 0 & 2 & 0 & 1 & 0.1 & 0 & -2.6 & 0.3 \\ 1.5 & 0.7 & 0.9 & 0 & 1 & 0 & 0.8 & 2 & 1 & -2.1 \end{bmatrix},$$

$$b^T = [2.2,\ 2.5,\ 1.5,\ 3.4,\ 1.1,\ 3.3,\ -1.9,\ 7.3,\ 2.3,\ 5.2].$$

Solve the following systems using method 2.

46. $$D = \begin{bmatrix} 1 & -0.2 & 0.1 \\ -0.1 & 1 & -0.1 \\ -0.3 & 0.2 & -1 \end{bmatrix},$$

$$C = \begin{bmatrix} 0.1 & 0 & 0 \\ 0 & 0.1 & 0 \\ 0 & 0 & 0.1 \end{bmatrix}, \quad b = \begin{bmatrix} 0.4 \\ 0.8 \\ 0.2 \end{bmatrix},$$

$k = 0(1)15.$

47. $$D = \begin{bmatrix} 1.342 & 0.432 & -0.599 & 0.202 \\ 0.202 & 1.342 & 0.432 & -0.599 \\ -0.599 & 0.202 & 1.342 & 0.432 \\ 0.432 & -0.599 & 0.202 & 1.342 \end{bmatrix},$$

$$C = \begin{bmatrix} 1 & 0 & 0 & 0 \\ 0 & 1 & 0 & 0 \\ 0 & 0 & 1 & 0 \\ 1 & 0 & 0 & 1 \end{bmatrix}, \quad b = \begin{bmatrix} 1.941 \\ -0.230 \\ -1.941 \\ 0.230 \end{bmatrix},$$

$k = 0(0.01)0.15.$

48. $$D = \begin{bmatrix} 1.45 & -2.42 & 1.58 & -2.36 & 1.43 \\ -2.49 & 1.63 & -2.94 & 1.99 & 1.88 \\ 0 & 1.79 & -2.59 & 1.26 & -2.55 \\ 0 & 0 & 1.22 & -2.03 & 1.31 \\ 0 & 0 & 0 & 1.11 & -2.09 \end{bmatrix},$$

$$C = \begin{bmatrix} 0.1 & 0 & 0 & 0 & 0 \\ 0 & 0.1 & 0 & 0 & 0 \\ 0 & 0 & 0.1 & 0 & 0 \\ 0 & 0 & 0 & 0.1 & 0 \\ 0 & 0 & 0 & 0 & 0.1 \end{bmatrix}, \; b = \begin{bmatrix} -1.20 \\ -2.33 \\ -0.14 \\ -0.09 \\ 2.19 \end{bmatrix},$$

$k = 0(1)14.$

49. $$D = \begin{bmatrix} 1.42 & 1.44 & 0 & 0 & 0 & 0 \\ 1.99 & 1.31 & -1.76 & 0 & 0 & 0 \\ 0 & 1.37 & 1.66 & 1.76 & 0 & 0 \\ 0 & 0 & -1.35 & 1.65 & -1.78 & 0 \\ 0 & 0 & 0 & 1.32 & 1.59 & 1.99 \\ 0 & 0 & 0 & 0 & -1.94 & 1.97 \end{bmatrix},$$

$$C = \begin{bmatrix} 0.1 & 0 & 0 & 0 & 0 & 0 \\ 0 & 0.1 & 0 & 0 & 0 & 0 \\ 0 & 0 & 0.1 & 0.1 & 0 & 0 \\ 0 & 0 & 0.1 & 0.1 & 0 & 0 \\ 0 & 0 & 0 & 0 & 0.1 & 0 \\ 0 & 0 & 0 & 0 & 0 & 0.1 \end{bmatrix}, \; b = \begin{bmatrix} 1.32 \\ 0.23 \\ 3.22 \\ 0.10 \\ 3.31 \\ 1.87 \end{bmatrix},$$

$k = 0(1)14.$

50. $$A = \begin{bmatrix} 2.12 & 0.42 & 1.34 & 0.88 & 0.75 & 1.21 & 0.39 & -2.11 \\ 1.34 & 0.95 & 1.87 & 0.43 & 0.54 & 1.25 & 0.75 & 0.91 \\ 0.88 & 1.87 & 0.98 & 0.46 & 1.71 & 0.41 & 1.87 & 0.19 \\ 0.75 & 0.43 & 0.46 & 2.44 & 1.14 & 0.36 & 1.13 & -2.54 \\ 1.21 & 0.54 & 1.71 & -2.14 & 1.49 & 1.71 & 0.39 & 0.87 \\ 0.39 & 1.25 & 0.41 & 0.36 & 1.71 & -1.51 & 0.42 & 1.14 \\ -2.11 & 0.75 & 1.87 & 1.13 & 0.39 & 0.42 & 1.91 & 0.45 \\ 1.52 & 0.91 & 0.19 & -2.54 & 0.87 & 1.14 & 0.45 & 2.35 \end{bmatrix},$$

$b^T = [4.4, \; 7.44, \; 7.77, \; 3.57, \; 5.18, \; 3.57, \; 4.21, \; 4.29].$

51. $$A = \begin{bmatrix} 0.422 & 0.511 & 0.633 & -0.744 & 0.322 & 0.211 & -0.823 & -0.901 & 0.102 \\ -0.511 & 0.211 & 0.366 & -0.477 & 0.255 & 0.712 & 0.656 & 0.544 & 0.122 \\ -0.633 & -0.366 & 0.501 & 0.411 & -0.734 & 0.266 & 0.133 & 0.812 & -0.956 \\ 0.744 & 0.477 & 0.411 & 0.355 & 0.144 & 0.978 & 0.155 & 0.788 & 0.611 \\ -0.320 & -0.322 & -0.255 & 0.734 & -0.144 & 0.166 & 0.277 & 0.000 & 0.733 \\ 0.455 & -0.923 & -0.211 & -0.711 & -0.266 & 0.978 & -0.277 & 0.191 & 0.566 \\ 0.755 & 0.233 & 0.823 & -0.656 & -0.133 & -0.155 & -0.734 & -0.566 & 0.222 \\ -0.533 & 0.111 & 0.901 & -0.544 & -0.812 & -0.788 & -0.455 & -0.755 & 0.533 \\ 0.555 & -0.844 & -1.101 & -0.122 & 0.956 & -0.611 & 0.922 & -0.233 & -0.111 \end{bmatrix}, \quad b = \begin{bmatrix} -0.267 \\ 1.878 \\ -0.566 \\ 4.663 \\ 0.721 \\ 1.258 \\ -1.621 \\ -2.209 \\ 1.677 \end{bmatrix},$$

52. $$A = \begin{bmatrix} 2.535 & -1.424 & 0.981 & 0.978 & -1.752 & 0.560 & -1.651 & 0.159 & 0.138 & 0.651 \\ 0.651 & 0.535 & 0.576 & 0.981 & 0.978 & -1.752 & 0.560 & -1.651 & 0.100 & 0.159 \\ 0.138 & 0.138 & 0.651 & 2.535 & -1.424 & 0.981 & 0.978 & -1.752 & 0.560 & -1.651 \\ 0.159 & 0.159 & 0.138 & 0.651 & 2.535 & -1.424 & 0.981 & 0.978 & -1.752 & 0.560 \\ -1.651 & -1.651 & 0.159 & 0.138 & 0.651 & 2.535 & -1.424 & 0.981 & 0.978 & -1.752 \\ 0.560 & 0.560 & -1.651 & 0.159 & 0.138 & 0.651 & 2.535 & -1.424 & 0.981 & 0.978 \\ -1.752 & -1.752 & 0.560 & -1.651 & 0.159 & 0.138 & 0.651 & 2.535 & -1.424 & 0.981 \\ 0.978 & 0.978 & -1.752 & 0.560 & -1.651 & 0.159 & -0.138 & 0.651 & 2.535 & -1.424 \\ 0.981 & 0.981 & 0.978 & -1.752 & 0.560 & -1.651 & 0.159 & 0.139 & 0.138 & 2.651 \\ 0.535 & -1.424 & -1.424 & 0.981 & 0.978 & -1.752 & 0.560 & -1.651 & 0.159 & 0.138 \end{bmatrix},$$

$$b^T = [-0.825, \ -0.825, \ -0.825, \ -0.825, \ -0.825, \ -0.825, \ -0.825, \ -0.825, \ -0.825, \ -0.825].$$

53. Using **2**, invert the matrices of the following problems and determine their condition number M:

a) Problem 46, $k = 0\,(1)\,5$.
b) Problem 47, $k = 0\,(0.01)\,0.1$.
c) Problem 48, $k = 0\,(1)\,7$.
d) Problem 49, $k = 0\,(1)\,5$.
e) Problem 50.
f) Problem 51.
g) Problem 52.

Use **3** to solve the following systems:

54. $$D = \begin{bmatrix} 2 & 1 & 1 \\ 1 & 2.5 & 1 \\ 1 & 1 & 3 \end{bmatrix}, \quad C = \begin{bmatrix} 1 & 0 & 0 \\ 0 & 1 & 0 \\ 0 & 0 & 1 \end{bmatrix}, \quad b = \begin{bmatrix} 4 \\ 4.5 \\ 5 \end{bmatrix},$$

$k = 0\,(0.1)\,1.5$.

55. $$D = \begin{bmatrix} 5 & 7 & 6 & 5 \\ 7 & 10 & 8 & 7 \\ 6 & 8 & 10 & 9 \\ 5 & 7 & 9 & 10 \end{bmatrix},$$

$$C = \begin{bmatrix} 0.1 & 0 & 0 & 0.1 \\ 0 & 0.1 & 0 & 0 \\ 0 & 0 & 0.1 & 0 \\ 0.1 & 0 & 0 & 0.1 \end{bmatrix}, \quad b = \begin{bmatrix} 23 \\ 32 \\ 33 \\ 31 \end{bmatrix},$$

$k = 0\,(0.1)\,1.5$.

56. $$D = \begin{bmatrix} 1.28 & 2.32 & 4.14 & -3.24 & -5.15 \\ 2.32 & 1.49 & 5.26 & 1.56 & 3.92 \\ 4.14 & 5.26 & 4.06 & 2.44 & 4.39 \\ -3.24 & 1.56 & 2.44 & 5.42 & 1.94 \\ -5.15 & 3.92 & 4.39 & 1.94 & 4.63 \end{bmatrix},$$

$$C = \begin{bmatrix} 0.1 & 0 & 0 & 0 & 0 \\ 0 & 0.1 & 0 & 0 & 0 \\ 0 & 0 & 0.1 & 0 & 0 \\ 0 & 0 & 0 & 0.1 & 0 \\ 0 & 0 & 0 & 0 & 0.1 \end{bmatrix}, \quad b = \begin{bmatrix} -3.02 \\ 3.26 \\ 0.83 \\ -8.20 \\ -6.45 \end{bmatrix},$$

$k = 0\,(0.1)\,1.5$.

57. $$D = \begin{bmatrix} 0.393 & 0.456 & 0.782 & -0.978 & 0.832 & 0.665 \\ 0.456 & 0.551 & 0.358 & 0.588 & 0.625 & 0.872 \\ 0.782 & 0.358 & 0.654 & 0.540 & 0.265 & 0.325 \\ -0.978 & 0.588 & 0.540 & 0.832 & 0.778 & 0.466 \\ 0.832 & 0.625 & 0.265 & 0.778 & 0.371 & 0.492 \\ 0.665 & 0.872 & 0.325 & 0.466 & 0.492 & 0.635 \end{bmatrix},$$

$$C = \begin{bmatrix} 0.1 & 0 & 0 & 0 & 0 & 0 \\ 0 & 0.1 & 0 & 0 & 0 & 0 \\ 0 & 0 & 0.1 & 0 & 0 & 0 \\ 0 & 0 & 0 & 0.1 & 0 & 0 \\ 0 & 0 & 0 & 0 & 0.1 & 0 \\ 0 & 0 & 0 & 0 & 0 & 0.1 \end{bmatrix}, \quad b = \begin{bmatrix} 0.510 \\ -0.058 \\ 1.111 \\ 0.904 \\ 0.605 \\ 0.355 \end{bmatrix},$$

$k = 0(1)\,15.$

58. $$A = \begin{bmatrix} 2.2 & 4 & -3 & 1.5 & 0.6 & 2 & 0.7 \\ 4 & 3.2 & 1.5 & -0.7 & -0.8 & 3 & 1 \\ -3 & 1.5 & 1.8 & 0.9 & 3 & 2 & 2 \\ 1.5 & -0.7 & 0.9 & 2.2 & 4 & 3 & 1 \\ 0.6 & -0.8 & 3 & 4 & 3.2 & 0.6 & 0.7 \\ 2 & 3 & 2 & 3 & 0.6 & 2.2 & 4 \\ 0.7 & 1 & 2 & 1 & 0.7 & 4 & 3.2 \end{bmatrix},$$

$b^T = [3.2,\ 4.3,\ -0.1,\ 3.5,\ 5.3,\ 9.0,\ 3.7].$

59. $$A = \begin{bmatrix} 3.4 & 8 & 5 & 4 & 2 & 1 & 6 & -7 \\ 8 & 1.4 & 4 & 3 & 8 & 5 & 2 & 7 \\ 5 & 4 & 2.4 & 4 & -5 & 1 & 3 & -8 \\ 4 & 3 & 4 & 1.4 & 6 & 7 & 2 & 1 \\ 2 & 8 & -5 & 6 & 3.4 & 8 & 4 & 5 \\ 1 & 5 & 1 & 7 & 8 & 2.4 & 3 & 6 \\ 6 & 2 & 3 & 2 & 4 & 3 & 1.4 & 7 \\ -7 & 7 & -8 & 1 & 5 & 6 & 7 & 3.4 \end{bmatrix},$$

$b^T = [22,\ 38,\ 6,\ 28,\ 31,\ 33,\ 28,\ 14].$

60. $$A = \begin{bmatrix} 4.1 & 2 & 3 & 1 & 1.6 & 2 & 1.5 & 0.7 & 1 \\ 2 & 5 & 1.4 & 0.5 & 0.7 & 3 & 0.8 & 2 & 3 \\ 3 & 1.4 & 3.8 & 1 & 2 & 1.4 & 3 & 1 & 0.8 \\ 1 & 0.5 & 1 & 5.1 & 0.8 & 1.5 & 2 & 1.6 & 0.5 \\ 1.6 & 0.7 & 2 & 0.8 & 4 & 0.9 & 0.7 & 3 & 2.4 \\ 2 & 3 & 1.4 & 1.5 & 0.9 & 5.3 & 3 & 2 & 2 \\ 1.5 & 0.8 & 3 & 2 & 0.7 & 3 & 4.1 & 1 & 3 \\ 0.7 & 2 & 1 & 1.6 & 3 & 2 & 1 & 5.5 & 2 \\ 1 & 3 & 0.8 & 0.5 & 2.4 & 2 & 3 & 2 & 4 \end{bmatrix},$$

$b^T = [16.9,\ 18.4,\ 17.4,\ 14.0,\ 16.1,\ 21.1,\ 19.1,\ 18.8,\ 18.0].$

61. $$A = \begin{bmatrix} 1.4481 & 0.5542 & 1.6602 & 0.8832 & 1.8061 & 0.4051 & 0.2918 & 1.4052 & 1.7263 & 0.9051 \\ 0.5542 & 2.7074 & -1.6370 & 1.9182 & 0.7263 & -1.7093 & 0.9145 & -1.7293 & 0.8091 & 0.6166 \\ 1.6602 & -1.6370 & 1.2226 & 1.1109 & 0.1713 & 1.1528 & 0.9261 & -1.7371 & 1.8122 & 0.5183 \\ 0.8832 & 1.9182 & 1.1109 & 2.8164 & 1.6142 & 0.5089 & -1.2407 & 0.6072 & 0.7153 & 1.1149 \\ 1.8061 & 0.7263 & 0.1713 & 1.6142 & 2.0109 & 0.8212 & 1.4117 & 0.2709 & 1.6142 & 1.3082 \\ 0.4051 & -1.7093 & 1.1528 & 0.5089 & 0.8212 & 2.2416 & 0.9112 & 0.5058 & 1.6212 & 0.6162 \\ 0.2918 & 0.9145 & 0.9261 & -1.2407 & 1.4117 & 0.9112 & 2.2147 & 0.6321 & 1.2304 & 0.6202 \\ 1.4052 & -1.7293 & -1.7371 & 0.6072 & 0.2709 & 0.5058 & 0.6321 & 2.1213 & 0.9723 & 1.6644 \\ 1.7263 & 0.8091 & 1.8122 & 0.7153 & 1.6142 & 1.6212 & 1.2304 & 0.9723 & 1.8351 & 1.9724 \\ 0.9051 & 0.6166 & 0.5183 & 1.1149 & 1.3082 & 0.6162 & 0.6202 & 1.6644 & 1.9724 & 1.7312 \end{bmatrix},$$

$$b^T = [10.0853,\ 2.1707,\ 4.2003,\ 9.0485,\ 10.7550,\ 6.0747,\ 6.9020,\ 3.6930,\ 13.3085,\ 10.0635].$$

Solve the following systems using 4. Taking the values determined as an initial approximation accurate to two significant places, use either 5 or 6 to improve the accuracy of the solution to five significant places.

62. $$D = \begin{bmatrix} 6.214 & 2.180 & 3.184 \\ -1.351 & 8.224 & 5.224 \\ 2.489 & -0.459 & 4.299 \end{bmatrix},$$

$$C = \begin{bmatrix} 0.5 & 0 & 0.5 \\ 0 & 0.5 & 0 \\ 0 & 0 & 0.5 \end{bmatrix}, \quad b = \begin{bmatrix} 49.91 \\ 50.17 \\ 32.68 \end{bmatrix},$$

$k = 0(0.5)\,7.5.$

63. $$D = \begin{bmatrix} 6.22 & 1.42 & -1.72 & 1.91 \\ 1.44 & 5.33 & 1.11 & -1.82 \\ -1.72 & 1.11 & 5.24 & 1.42 \\ 1.91 & -1.82 & 1.42 & 6.55 \end{bmatrix},$$

$$C = \begin{bmatrix} 1 & 0 & 0 & 0 \\ 0 & 1 & 0 & 0 \\ 0 & 0 & 1 & 0 \\ 0 & 0 & 0 & 1 \end{bmatrix}, \quad b = \begin{bmatrix} 7.53 \\ 6.06 \\ 8.05 \\ 8.06 \end{bmatrix},$$

$k = 0(1)\,15.$

64. $$D = \begin{bmatrix} 6.2 & 0.1 & 0.3 & 0.4 & -0.5 \\ 0.5 & 5.4 & 0.1 & 0.2 & 0.4 \\ 0 & 0.4 & 6.3 & 0.6 & 0.2 \\ 0 & 0 & 0.4 & 5.3 & 0.5 \\ 0 & 0 & 0 & 0.1 & 6.2 \end{bmatrix},$$

$$C = \begin{bmatrix} 1 & 0 & 0 & 0 & 0 \\ 0 & 1 & 0 & 0 & 0 \\ 0 & 0 & 1 & 0 & 0 \\ 0 & 0 & 0 & 1 & 0 \\ 0 & 0 & 0 & 0 & 1 \end{bmatrix}, \quad b = \begin{bmatrix} 0.1 \\ 0.4 \\ -0.3 \\ 1.1 \\ -0.1 \end{bmatrix},$$

$k = 0(1)\,15.$

65. $$D = \begin{bmatrix} 4.324 & 0.625 & 0.541 & 0.932 & -0.244 & 0.877 \\ 0.774 & 4.562 & 0.423 & 0.392 & 0.442 & 0.787 \\ 0.000 & -0.256 & 3.732 & 0.623 & 0.562 & 0.326 \\ 0.000 & 0.000 & -0.265 & 4.570 & 0.351 & -0.237 \\ 0.425 & 0.721 & 0.000 & 0.725 & 4.831 & 0.675 \\ 0.000 & 0.172 & 0.392 & -0.413 & 0.385 & 4.565 \end{bmatrix}$$

$$C = \begin{bmatrix} 1 & 0 & 0 & 0 & 0 & 0 \\ 0 & 1 & 0 & 0 & 0 & 0 \\ 0 & 0 & 1 & 0 & 0 & 0 \\ 0 & 0 & 0 & 1 & 0 & 0 \\ 0 & 0 & 0 & 0 & 1 & 0 \\ 0 & 0 & 0 & 0 & 0 & 1 \end{bmatrix}, \quad b = \begin{bmatrix} -0.272 \\ 0.746 \\ -0.029 \\ 0.542 \\ 0.375 \\ 0.586 \end{bmatrix},$$

$k = 0(1)\,15.$

66. $$A = \begin{bmatrix} 15.21 & 0 & 1.11 & 2.13 & 1.14 & 0.98 & 0.44 \\ 1.32 & 14.82 & 0.61 & 1.42 & 0.74 & 0.34 & 1.33 \\ 0.75 & 1.26 & 15.44 & 0.22 & 1.56 & 0.66 & 0.44 \\ 1.35 & 1.42 & 0.33 & 14.32 & 2.11 & 0.23 & 1.12 \\ 0.81 & 0.79 & 1.38 & 0.82 & 15.99 & 0.89 & 0.83 \\ 1.49 & 0.57 & 0.32 & 1.49 & 0.81 & 14.93 & 0.97 \\ 1.18 & 0.42 & 1.35 & 0.19 & 1.26 & 0.91 & 15.71 \end{bmatrix},$$

$b^T = [9.01,\ 8.58,\ 8.33,\ 8.88,\ 9.51,\ 8.58,\ 9.02].$

67. $$A = \begin{bmatrix} 11.343 & 0.522 & 0.451 & 0.842 & -0.971 & 0.875 & 0 & 0.175 \\ 0.245 & 13.378 & 0 & 0.356 & 0.625 & -0.910 & 0.262 & 0.217 \\ 0 & 0.394 & 18.82 & -0.911 & 0 & 0.429 & 0.358 & 0.232 \\ 0.871 & 0.377 & 0.423 & 10.707 & 0.625 & 0.563 & 0 & 0.671 \\ 0.378 & 0.921 & 0.417 & 0.832 & 9.322 & 0.444 & 0.812 & 0.542 \\ 0.629 & 0.458 & 0 & 0.732 & -0.982 & 9.254 & 0.587 & 0.391 \\ 0 & 0.494 & 0.358 & 0.675 & 0.415 & 0.604 & 10.804 & -0.908 \\ 0.422 & 0.155 & 0.532 & 0 & 0.771 & 0.475 & 0.415 & 9.108 \end{bmatrix},$$

$b^T = [2.237,\ 1.173,\ 1.384,\ 4.237,\ 4.668,\ 2.069,\ 2.442,\ 2.878].$

68. $$A = \begin{bmatrix} 16.91 & 0 & 0.32 & 1.42 & 0.44 & 0.64 & 1.19 & 0 & 1.71 \\ 0.32 & 14.79 & 0.99 & 1.11 & 0.49 & -1.18 & 0.23 & 0.65 & 0.48 \\ 1.21 & 0 & 16.52 & 0.29 & -1.44 & 0 & 1.26 & 0.81 & -1.39 \\ 0.53 & 0.85 & 1.24 & 16.34 & -1.15 & 0.84 & 1.17 & 0.31 & 0 \\ 0.43 & 0.20 & 0.21 & 1.25 & 16.36 & 1.12 & 0.91 & 1.16 & 0.52 \\ -1.13 & 0.42 & 0.52 & -1.45 & 0 & 16.51 & 0.49 & 1.31 & 0.32 \\ 1.33 & 0.62 & -1.44 & 0.24 & 0.65 & 0.36 & 16.00 & -1.19 & 1.16 \\ 0.77 & 0 & 0.81 & 1.31 & 0.29 & 1.12 & 0.43 & 16.45 & 0 \\ 1.23 & 1.57 & 0.98 & 0 & 1.12 & 0.75 & 0.55 & 1.21 & 16.34 \end{bmatrix},$$

$b^T = [6.63,\ 1.88,\ 1.26,\ 4.13,\ 4.83,\ 4.45,\ 1.17,\ 5.64,\ 7.65].$

$$
\textbf{69. } A = \begin{bmatrix}
17.3 & -2.3 & 0.5 & 0 & 0 & 0.4 & 0.6 & 0 & 1.7 & 0 \\
0.9 & 14.2 & 1.5 & 0 & 0.6 & 0.9 & 0 & 0 & 0.4 & 0 \\
0 & -2.1 & 16.7 & 1.4 & 0.3 & 0 & 0.1 & 0 & 0.8 & 0.3 \\
2 & 1 & 1.2 & 16.4 & -2.2 & 0 & 0.7 & 0.2 & 0 & 0.1 \\
0 & 0.8 & 0 & -2.8 & 16.7 & 1.6 & 0.5 & 0 & 0.3 & 0 \\
0.6 & 0.4 & 1 & 0 & 1.5 & 16.9 & -2.5 & 0.6 & 0 & 0.8 \\
0 & 2 & 0 & 0.2 & 0 & -2.1 & 16.3 & -2.3 & 0 & 0 \\
0.5 & 0 & 0.7 & 0.1 & 0.9 & 0.4 & 0.6 & 17.7 & 1.4 & 1 \\
0 & 0 & 2 & 0 & 1 & 0.1 & 0 & -2.6 & 16.3 & 1.5 \\
0.7 & 0.9 & 0 & 1 & 0 & 0.8 & 2 & 1 & -2.1 & 16.0
\end{bmatrix},
$$

$$b^T = [2.2,\ 2.5,\ 1.5,\ 3.4,\ 1.1,\ 3.3,\ -1.9,\ 7.3,\ 2.3,\ 5.2].$$

2. The calculation of eigenvalues and eigenvectors of matrices

THEORY

Methods for the determination of eigenvalues and eigenvectors of matrices can be divided into two basically distinct groups: (1) exact and (2) iterative.

In the methods of (1), the exact values of the coefficients p_i of the characteristic equation (b) are determined (if the values of the elements of A and the calculations are exact). The eigenvalues λ_i are then obtained as the roots of the n-th order polynomial (b). The corresponding eigenvectors are determined directly, without solving homogeneous systems of type (c).

In iterative methods, the eigenvalues and eigenvectors are determined directly, without constructing (b), and are the limits of numerical sequences.

The methods of (1) and (2) include those which yield all the eigenvalues and eigenvectors and those which only give one, or a few, of them.

Exact methods)*

10. Krylov's method

Let c_0 be an arbitrary vector, for example $c_0^T = (1, 0, \ldots, 0)$, and define the vectors c_i $(i = 1, 2, \ldots, n)$ inductively by

$$c_i = Ac_{i-1}.$$

*) see [4; chapters 6–8], [5; chapter 4], [6; chapter 10] and [3; chapter 6].

The coefficients p_i of (b) are the solutions of the linear algebraic system

$$c_n = c_{n-1} p_1 + c_{n-2} p_2 + \cdots, + c_1 p_{n-1} + c_0 p_n. \tag{d}$$

If the values of the c_i, p_i and λ_i are known, the eigenvectors x_i are defined by

$$x_i = \sum_{j=0}^{n-1} a_{ij} c_{n-1-j},$$

where

$$a_{i0} = 1, \quad a_{ij} = \lambda_i a_{i,j-1} - p_j \quad (j = 1, 2, \ldots, n-1).$$

Note. It is not always possible to determine the coefficients of (b) by Krylov's method. If c_0 is not chosen carefully, the minimal equation {the roots of which coincide with those of (b) but have lower multiplicity} or one of its divisors may be obtained instead of (b).

These exceptional cases occur when the rows of (a) are not linearly independent.

11. Danilevskiy's method

In this method, $n - 1$ transformations are used to construct the sequence $A = A^{(1)}, A^{(2)}, \ldots, A^{(n)} = B$ so that B is a Frobenius matrix

$$B = \begin{bmatrix} 0 & 0 & \ldots & 0 & p_n \\ 1 & 0 & \ldots & 0 & p_{n-1} \\ 0 & 1 & \ldots & 0 & p_{n-2} \\ \cdot & \cdot & \cdot & \cdot & \cdot \\ 0 & 0 & \ldots & 1 & p_1 \end{bmatrix}$$

with the elements p_i corresponding to the coefficients of (b). Writing $[A^{(k)}]_{ij} = a_{ij}^{(k)}$ $(k = 1, 2, \ldots, n)$, the process is defined inductively by

$$b_{k+1j}^{(k)} = \frac{1}{a_{k+1k}^{(k)}} \cdot a_{k+1j}^{(k)},$$

$$b_{ij}^{(k)} = a_{ij}^{(k)} - a_{ik}^{(k)} b_{k+1j}^{(k)} \qquad (i \neq k + 1),$$

$$a_{ij}^{(k+1)} = b_{ij}^{(k)} \qquad (j \neq k + 1),$$

$$a_{i\,k+1}^{(k+1)} = \sum_{j=1}^{n} b_{ij}^{(k)} a_{jk}^{(k)}.$$

The stability of this inductive process depends heavily on the magnitudes of the $a^{(k)}_{k+1,k}$. For the difficult case when $a^{(k)}_{k+1,k} \doteq 0$ for some k, see [5; § 46], [7; chapters 6 and 7] and [8].

The components of the eigenvector $x = (x_1, x_2, \ldots, x_n)^T$, corresponding to the eigenvalue λ, are calculated using the recurrence relations

$$y_i = \lambda^{n-i} - \sum_{j=1}^{n-i} p_j \lambda^{n-i-j} \qquad (i = n, n-1, \ldots, 1),$$

$$\left.\begin{aligned} y_i^{(k)} &= y_i^{(k+1)} + y_{k+1}^{(k+1)} a_{ik}^{(k)} \qquad (i \neq k+1), \\ y_{k+1}^{(k)} &= y_{k+1}^{(k+1)} a_{k+1k}^{(k)}, \end{aligned}\right\}$$

with

$$y_i = y_i^{(n)}, \quad y_i^{(1)} = x_i \qquad (k = n-1, n-2, \ldots, 1).$$

12. Faddeev's modification of Leverrier's method (see [5; § 47])

The basic steps in Leverrier's method are

1. Compute recursively the powers of the matrix A

$$A^k = A^{k-1} A \qquad (k = 1, 2, \ldots, n);$$

2. Calculate the traces $tr\,A^k$

$$tr\,A^k = \lambda_1^k + \lambda_2^k + \cdots + \lambda_n^k = \sum_{i=1}^{n} a_{ii}^{(k)} \; (A^k = [a_{ij}^{(k)}]);$$

3. The coefficients of (b) are then defined recursively by

$$k\,p_k = tr\,A^k - p_1\,tr\,A^{k-1} - \cdots - p_{k-1}\,tr\,A.$$

Using Faddeev's modification, it is possible not only to find the coefficients of (b) but also to determine the inverse matrix and the eigenvectors. The sequence $A_1, A_2, \ldots, A_n$ is constructed such that

$$A_1 = A, \quad tr\,A_1 = p_1, \quad C_1 = A_1 - p_1 I,$$

$$A_2 = AC_1, \quad \frac{tr\,A_2}{2} = p_2, \quad C_2 = A_2 - p_2 I,$$

$$\cdots\cdots\cdots\cdots\cdots\cdots\cdots\cdots$$

$$A_{n-1} = AC_{n-2}, \quad \frac{tr\,A_{n-1}}{n-1} = p_{n-1}, \quad C_{n-1} = A_{n-1} - p_{n-1} I,$$

$$A_n = AC_{n-1}, \quad \frac{tr\,A_n}{n} = p_n, \quad C_n = A_n - p_n I.$$

If A is non-singular, then

$$A^{-1} = \frac{C_{n-1}}{p_n}.$$

If the eigenvalues λ_k are distinct, then the corresponding eigenvectors x_k are defined by the recurrence relation

$$\bar{x}_0 = e, \quad \bar{x}_i = \lambda_k \bar{x}_{i-1} + c_i, \quad \bar{x}_0 = e, \quad \bar{x}_i = \lambda_k \bar{x}_{i-1} + c_i,$$
$$x_k = \bar{x}_{n-1},$$

where c_i is any column of the matrix C_i and e is the corresponding column of the unit matrix.

13. The interpolation method

Let

$$\alpha_0 = 0, \quad \alpha_1 = 1, \ldots, \alpha_n = n$$

or, more generally,

$$\alpha_i = a + h\,i \qquad (i = 0, 1, \ldots, n),$$

and compute

$$D(\alpha_i) = \det (A - \alpha_i I).$$

The characteristic polynomial

$$D(\lambda) = \det (A - \lambda I)$$

is then constructed using

$$D(\lambda) = D(\alpha_0) + \sum_{i=1}^{n} \frac{\Delta^i D(\alpha_0)}{i!} \lambda(\lambda - 1) \cdots (\lambda - i + 1),$$

where Δ denotes the forward difference operator {see [6; § 3.3–2–2, Fig. 3.1]}, viz.

$$\Delta D(\alpha_j) = D(\alpha_j + h) - D(\alpha_j).$$

Iterative methods)*

14. The classical method of Jacobi

This method will give all the eigenvalues and eigenvectors of any *symmetric* matrix A. The matrix A is transformed to diagonal

*) see [4; chapter 9] and [5; chapter 5 and 8].

form by a series of similarity transformations. Using the transformation $A^{(k+1)} = T_{ij}^{(k)} A^{(k)} (T_{ij}^{(k)})^T$ ($k = 0, 1, \ldots$) with

$$T_{ij}^{(k)} = \begin{bmatrix} 1 & & & & & & \\ & \ddots & & & & & \\ & & c & \cdots & -s & & \\ & & \vdots & 1 & \vdots & & \\ & & s & \cdots & c & & \\ & & & & & \ddots & \\ & & & & & & 1 \end{bmatrix} \begin{matrix} \\ \\ \cdots i \\ \\ \cdots j \\ \\ \\ \end{matrix} \qquad (i < j),$$

the sequence $A^{(0)} = A, A^{(1)}, A^{(2)}, \ldots,$ is constructed.
The elements c and s are chosen so that the off-diagonal elements $a_{ij}^{(k)}$ and $a_{ji}^{(k)}$ of $A^{(k)}$ are eliminated. This elimination of off-diagonal elements is arranged so that the elements of maximum modulus of the remaining off-diagonal elements are always eliminated next. On obtaining $A^{(k)}$, if $a_{ji}^{(k)}$ is its element of maximum modulus, the process is continued inductively by defining c and s of $T_{ij}^{(k)}$ so that

$$c = \sqrt{\frac{1}{2}\left(1 + \frac{|a_{ii}^{(k)} - a_{jj}^{(k)}|}{d}\right)};$$

$$d = \sqrt{(a_{ii}^{(k)} - a_{jj}^{(k)})^2 + 4\,(a_{ij}^{(k)})^2};$$

$$s = \operatorname{sign}\,[a_{ij}^{(k)}(a_{ii}^{(k)} - a_{jj}^{(k)})]\sqrt{\frac{1}{2}\left(1 - \frac{|a_{ii}^{(k)} - a_{jj}^{(k)}|}{d}\right)}.$$

The i-th and j-th columns of $A^{(k)}$ are replaced by

$$\begin{aligned} a_{ti}^{(k+1)} &= c\,a_{ti}^{(k)} + s\,a_{tj}^{(k)}, \\ a_{tj}^{(k+1)} &= -\,s\,a_{ti}^{(k)} + c\,a_{tj}^{(k)} \qquad (t = 1, 2, \ldots, n), \end{aligned}$$

while its i-th and j-th rows become

$$\begin{aligned} a_{it}^{(k+1)} &= c\,a_{it}^{(k)} + s\,a_{jt}^{(k)}, \\ a_{jt}^{(k+1)} &= -\,s\,a_{it}^{(k)} + c\,a_{jt}^{(k)} \qquad (t = 1, 2, \ldots, n), \\ a_{ij}^{(k+1)} &= a_{ji}^{(k+1)} = 0. \end{aligned}$$

The process continues until all of the off-diagonal elements are less than some given threshold ε. Then

$$\lambda_i \approx a_{ii}^{(k+1)}.$$

If the eigenvalues of A are simple and the off-diagonal elements do not exceed some small number ε, then the components of the eigenvectors x_i can be determined with an accuracy of ε^2 by using

$$x_i = \left(\prod_{k=1}^{\nu} T_{i_k j_k}\right) v_i,$$

$$v_i = \left(\frac{\alpha_{1i}}{\alpha_{ii} - \alpha_{11}}, \ldots, \frac{\alpha_{i-1,i}}{\alpha_{ii} - \alpha_{i-1,i-1}}, 1, \frac{\alpha_{i+1,i}}{\alpha_{ii} - \alpha_{i+1,i+1}}, \ldots, \frac{\alpha_{ni}}{\alpha_{ii} - \alpha_{nn}}\right),$$

where the α_{ij} are the elements of $A^{(k)}$ with which the transformation process terminated.

15. Matrix powering

This method can be used to determine the maximum modulus eigenvalue and corresponding eigenvector of a given matrix A. Let λ_1, the maximum modulus eigenvalue of A, be real and simple. Starting with an arbitrary vector $y^{(0)}$, the following sequence of vectors is constructed:

$$\begin{aligned} y^{(1)} &= Ay^{(0)}, \\ y^{(2)} &= A^2 y^{(0)} = A y^{(1)}, \\ &\cdots\cdots\cdots\cdots \\ y^{(k)} &= A^k y^{(0)} = Ay^{(k-1)}, \\ &\cdots\cdots\cdots\cdots \end{aligned}$$

For sufficiently large k, the ratio of corresponding components of $y^{(k+1)}$ and $y^{(k)}$ is approximately equal to the required eigenvalue λ_1,

$$\lambda_1 \approx \frac{y_i^{(k+1)}}{y_i^{(k)}} \qquad (i = 1, 2, \ldots, n). \tag{e}$$

The eigenvector corresponding to λ_1 is also determined by this method: the ratios of the components of $y^{(k)}$ tend to the ratios of the components of the required eigenvector.

If A is symmetric, then, for sufficiently large k,

$$\lambda_1 \approx (y^{(k+1)})^T y^{(k+1)} / (y^{(k)})^T y^{(k+1)}.$$

Using this expression instead of (e), the rate of convergence of the approximation to λ_1 is almost doubled.

PROBLEMS

In examples 70–146, use the method suggested to determine the eigenvalues and eigenvectors of either the matrix A or $A = D + kC$ for the values of k specified.

Using Krylov's method, determine the eigenvalues and eigenvectors of the following matrices:

70. $$A = \begin{bmatrix} 0.012282 & -0.183120 & 0.073265 & 0.773476 \\ -0.183120 & -0.084200 & -0.761983 & -0.043694 \\ 0.073265 & -0.761983 & 0.110212 & -0.202017 \\ 0.773476 & -0.043694 & -0.202017 & -0.244178 \end{bmatrix}.$$

71. $$A = \begin{bmatrix} -0.259222 & -0.492228 & 0.072616 & 0.629063 \\ -0.492228 & 0.199925 & -0.604859 & 0.026396 \\ 0.072616 & -0.604859 & -0.155300 & -0.471768 \\ 0.629063 & 0.026396 & -0.471768 & 0.036554 \end{bmatrix}.$$

72. $$A = \begin{bmatrix} -0.349314 & 0.634757 & -0.050233 & 0.476204 \\ 0.634757 & 0.237876 & -0.480047 & -0.065226 \\ -0.050233 & -0.480047 & -0.218524 & 0.576976 \\ 0.476204 & -0.065226 & 0.576976 & 0.141277 \end{bmatrix}.$$

73. $$A = \begin{bmatrix} -0.335584 & -0.741744 & 0.060428 & 0.309504 \\ -0.741744 & 0.244031 & -0.316031 & 0.066983 \\ 0.060428 & -0.316031 & -0.233649 & -0.642471 \\ 0.309504 & 0.066983 & -0.642471 & -0.160941 \end{bmatrix}.$$

74. $$D = \begin{bmatrix} 0.223 & 0.231 & 0.412 \\ 0.231 & 0.556 & 0.564 \\ 0.412 & 0.564 & 0.889 \end{bmatrix}, \quad C = \begin{bmatrix} 1 & 0 & 1 \\ 0 & 1 & 0 \\ 1 & 0 & 1 \end{bmatrix},$$

$k = 0(0.1)\,0.6.$

75. $$D = \begin{bmatrix} 2 & 1 & 1 \\ 1 & 2.5 & 1 \\ 1 & 1 & 3 \end{bmatrix}, \quad C = \begin{bmatrix} 1 & 0 & 1 \\ 0 & 1 & 1 \\ 1 & 1 & 1 \end{bmatrix},$$

$k = 0(0.1)1.5.$

76. $D = \begin{bmatrix} 1 & 1.5 & 2.5 & 3.5 \\ 1.5 & 1 & 2 & 1.6 \\ 2.5 & 2 & 1 & 1.7 \\ 3.5 & 1.6 & 1.7 & 1 \end{bmatrix}, \quad C = \begin{bmatrix} 1 & 0 & 1 & 1 \\ 0 & 1 & 0 & 0 \\ 1 & 0 & 1 & 0 \\ 1 & 0 & 0 & 1 \end{bmatrix},$

$k = 0(1)\,15.$

77. $D = \begin{bmatrix} 9 & 8 & 13 & -3 \\ 8 & 9 & 6 & 11 \\ 13 & 6 & 3 & 10 \\ -3 & 11 & 10 & 9 \end{bmatrix}, \quad C = \begin{bmatrix} 1 & 0 & 1 & 1 \\ 0 & 1 & 0 & 1 \\ 1 & 0 & 0 & 1 \\ 1 & 1 & 1 & 1 \end{bmatrix},$

$k = 0(1)\,4.$

78. $D = \begin{bmatrix} 1 & 5 & 7 & 3 & 9 \\ 5 & 0 & 6 & -4 & 3 \\ 7 & 6 & -5 & 7 & 2 \\ 3 & -4 & 7 & -1 & 4 \\ 9 & 3 & 2 & 4 & 0 \end{bmatrix}, \quad C = \begin{bmatrix} 0.1 & 0 & 0.1 & 0 & 0.1 \\ 0 & 0.1 & 0 & 0.1 & 0 \\ 0.1 & 0 & 0.1 & 0 & 0 \\ 0 & 0.1 & 0 & 0.1 & 0 \\ 0.1 & 0 & 0 & 0 & 0.1 \end{bmatrix},$

$k = 0(1)\,10.$

79. $D = \begin{bmatrix} 1.9 & 0.5 & -1.2 & 0.6 & 0.7 & 1.6 \\ 0.5 & 1.3 & 0.9 & -1.8 & 1.2 & 0.4 \\ -1.2 & 0.9 & 1.8 & 0.4 & 0.6 & 1.4 \\ 0.6 & -1.8 & 0.4 & 1.9 & 0.2 & -1.6 \\ 0.7 & 1.2 & 0.6 & 0.2 & 0.5 & 1.5 \\ 1.6 & 0.4 & 1.4 & -1.6 & 1.5 & 1.3 \end{bmatrix},$

$C = \begin{bmatrix} 0.5 & 0 & 0 & 0 & 0 & 0.5 \\ 0 & 0.5 & 0 & 0 & 0 & 0 \\ 0 & 0 & 0.5 & 0 & 0 & 0 \\ 0 & 0 & 0 & 0.5 & 0 & 0 \\ 0 & 0 & 0 & 0 & 0.5 & 0 \\ 0.5 & 0 & 0 & 0 & 0 & 0.5 \end{bmatrix},$

$k = 0(1)\,5.$

80. $D = \begin{bmatrix} 11 & 0.8 & 1.7 & 0.3 & 1.5 & 0.2 & 3.6 \\ 0.8 & 6.2 & 0.5 & 1.3 & 0.7 & 0.8 & 1.4 \\ 1.7 & 0.5 & 11.1 & 4.9 & 1.8 & -2.2 & -1.7 \\ 0.3 & 1.3 & 4.9 & 5.7 & 9.3 & 1.5 & 2.3 \\ 1.5 & 0.7 & 1.8 & 9.3 & 3.4 & 0.5 & 1.6 \\ 0.2 & 0.8 & -2.2 & 1.5 & 0.5 & 5.1 & 2.4 \\ 3.6 & 1.4 & -1.7 & 2.3 & 1.6 & 2.4 & 9.5 \end{bmatrix},$

$$C = \begin{bmatrix} 1 & 0 & 0 & 0 & 0 & 0 & 1 \\ 0 & 1 & 0 & 0 & 0 & 0 & 0 \\ 0 & 0 & 1 & 0 & 0 & 0 & 0 \\ 0 & 0 & 0 & 0 & 1 & 0 & 0 \\ 0 & 0 & 0 & 1 & 0 & 0 & 0 \\ 0 & 0 & 0 & 0 & 0 & 0 & 0 \\ 1 & 0 & 0 & 0 & 0 & 0 & 1 \end{bmatrix},$$

$k = 0(1)\,3.$

Using Krylov's method, calculate the eigenvalues of the following matrices:

81. $A = \begin{bmatrix} 2.1 & -4.5 & -2 \\ 3 & 2.5 & 4.3 \\ -6 & 3.5 & 2.5 \end{bmatrix}.$

82. $A = \begin{bmatrix} 1 & -2 & 0.4 \\ 3 & 0.36 & -0.5 \\ 2 & 1 & 1 \end{bmatrix}.$

83. $A = \begin{bmatrix} 1 & 7 & 6 & 5 \\ 7 & 10 & 8 & 7 \\ 6 & 8 & 10 & 9 \\ 5 & 7 & 9 & 10 \end{bmatrix}.$

84. $A = \begin{bmatrix} 1 & 0.2 & 0.04 & 0.008 \\ 1 & -0.6 & 0.36 & -0.216 \\ 1 & 0.8 & 0.64 & 0.512 \\ 1 & -0.4 & 0.16 & -0.064 \end{bmatrix}.$

Determine the eigenvalues and corresponding eigenvectors of the following matrices using Danilevskiy's method:

85. $A = \begin{bmatrix} 0.255123 & 0.822450 & -0.064280 & 0.184773 \\ 0.822450 & 0.181031 & -0.140328 & -0.048661 \\ -0.064280 & -0.140328 & -0.220208 & 0.695908 \\ 0.184773 & -0.048661 & 0.695908 & 0.104195 \end{bmatrix}.$

86. $A = \begin{bmatrix} 0.346915 & -0.877791 & 0.033625 & 0.010275 \\ -0.877791 & 0.037104 & -0.081546 & 0.042925 \\ 0.033625 & -0.081546 & -0.093404 & -0.743226 \\ 0.010275 & 0.042925 & -0.743226 & -0.254397 \end{bmatrix}.$

87. $A = \begin{bmatrix} 0.184738 & -0.823333 & 0.016842 & -0.086509 \\ -0.823333 & -0.299067 & -0.040318 & -0.021202 \\ 0.016842 & -0.040318 & 0.178882 & -0.715463 \\ -0.086509 & -0.021202 & -0.715463 & -0.241781 \end{bmatrix}.$

88. $A = \begin{bmatrix} 0.472200 & -0.634640 & 0.042951 & 0.083298 \\ -0.634640 & -0.537618 & -0.220341 & -0.074237 \\ 0.042951 & -0.220341 & 0.400799 & -0.600256 \\ 0.083298 & -0.074237 & -0.600256 & -0.539341 \end{bmatrix}.$

89. $D = \begin{bmatrix} 1.3 & 0.4 & 0.5 \\ 0.4 & 1.3 & 0.3 \\ 0.5 & 0.3 & 1.3 \end{bmatrix}, C = \begin{bmatrix} 1 & 1 & 1 \\ 1 & 1 & 0 \\ 1 & 0 & 1 \end{bmatrix},$

$k = 0(0.1)0.5.$

90. $D = \begin{bmatrix} 11.1 & 44.4 & 55.5 \\ 44.4 & 22.2 & 44.4 \\ 55.5 & 44.4 & 33.3 \end{bmatrix}, C = \begin{bmatrix} 1 & 0 & 1 \\ 0 & 1 & 0 \\ 1 & 0 & 1 \end{bmatrix},$

$k = 0(1)13.$

91. $D = \begin{bmatrix} 1.5 & 1.6 & 1.7 & 1.8 \\ 1.6 & 2.5 & 1.2 & 1.3 \\ 1.7 & 1.2 & 3.5 & 1.4 \\ 1.8 & 1.3 & 1.4 & 4.5 \end{bmatrix}, C = \begin{bmatrix} 0.1 & 0 & 0 & 0 \\ 0 & 0.1 & 0.1 & 0 \\ 0 & 0.1 & 0.1 & 0 \\ 0 & 0 & 0 & 0.1 \end{bmatrix},$

$k = 0(1)5.$

92. $D = \begin{bmatrix} 8.72 & 3.45 & 5.38 & 4.21 \\ 3.45 & 3.91 & 0.89 & 3.24 \\ 5.38 & 0.89 & 5.76 & 1.38 \\ 4.21 & 3.24 & 1.38 & 0.97 \end{bmatrix}, C = \begin{bmatrix} 0.5 & 0 & 1 & 1 \\ 0 & 0.5 & 0 & 0 \\ 1 & 0 & 0.5 & 0 \\ 1 & 0 & 0 & 0.5 \end{bmatrix},$

$k = 0(0.1)1.5.$

93. $D = \begin{bmatrix} 2.1 & 0.8 & 0.2 & 1.5 & 1 \\ 0.8 & 1.0 & 2.3 & 6.6 & 3 \\ 0.2 & 2.3 & 1.4 & 2.8 & 1 \\ 1.5 & 6.6 & 2.8 & 3.1 & 2 \\ 1 & 3 & 1 & 2 & 1 \end{bmatrix}, C = \begin{bmatrix} 1 & 0 & 0 & 0 & 1 \\ 0 & 1 & 1 & 1 & 0 \\ 0 & 1 & 1 & 1 & 0 \\ 0 & 1 & 1 & 1 & 0 \\ 1 & 0 & 0 & 0 & 1 \end{bmatrix},$

$k = 0(1)10.$

94. $D = \begin{bmatrix} 2.6 & -7.4 & 3 & 5 & 1.6 & 0.6 \\ -7.4 & 4.6 & 2 & 1 & 3.6 & -6.4 \\ 3 & 2 & 1.6 & 4 & 5 & 6 \\ 5 & 1 & 4 & 8.6 & 3 & 7 \\ 1.6 & 3.6 & 5 & 3 & 0.6 & 4 \\ 0.6 & -6.4 & 6 & 7 & 4 & 5.6 \end{bmatrix},$

$$C = \begin{bmatrix} 0.2 & 0.2 & 0 & 0 & 0.2 & 0.2 \\ 0.2 & 0.2 & 0 & 0 & 0.2 & 0.2 \\ 0 & 0 & 0.2 & 0 & 0 & 0 \\ 0 & 0 & 0 & 0.2 & 0 & 0 \\ 0.2 & 0.2 & 0 & 0 & 0.2 & 0 \\ 0.2 & 0.2 & 0 & 0 & 0 & 0.2 \end{bmatrix},$$

$k = 0(1)5.$

95. $$D = \begin{bmatrix} 2 & 0.8 & 1.7 & 0.3 & 1.5 & 0.2 & -1.4 \\ 0.8 & 1.2 & 0.5 & 1.3 & 0.7 & 0.8 & 1.4 \\ 1.7 & 0.5 & 2.1 & 0.9 & 1.8 & -2.2 & -1.7 \\ 0.3 & 1.3 & 0.9 & 1.7 & 0.3 & 1.5 & 2.3 \\ 1.5 & 0.7 & 1.8 & 0.3 & -1.4 & 0.5 & 1.6 \\ 0.2 & 0.8 & -2.2 & 1.5 & 0.5 & 1.1 & 2.4 \\ -1.4 & 1.4 & -1.7 & 2.3 & 1.6 & 2.4 & 0.5 \end{bmatrix},$$

$$C = \begin{bmatrix} 1 & 0 & 0 & 0 & 0 & 0 & 0 \\ 0 & 1 & 0 & 0 & 0 & 0 & 0 \\ 0 & 0 & 1 & 1 & 0 & 0 & 0 \\ 0 & 0 & 1 & 1 & 1 & 0 & 0 \\ 0 & 0 & 0 & 1 & 1 & 0 & 0 \\ 0 & 0 & 0 & 0 & 0 & 1 & 0 \\ 0 & 0 & 0 & 0 & 0 & 0 & 1 \end{bmatrix},$$

$k = 0(1)3.$

96. Use Danilevskiy's method to construct the characteristic equation for the matrices

$$D = \begin{bmatrix} 0.9 & 0 & 0 & 0.5 \\ 0 & 0.5 & 0.4 & 0 \\ 0 & 0.3 & 0 & 8 \\ 0 & 0.3 & 0 & 1 \end{bmatrix}, \quad C = \begin{bmatrix} 1 & 0 & 0 & 0 \\ 0 & 0.6 & 0.8 & 0 \\ 0 & -0.8 & 0.6 & 0 \\ 0 & 0 & 0 & 0 \end{bmatrix},$$

$k = 0(0.1)1.$

Construct the characteristic equation for the following matrices using Leverrier's method:

97. $$A = \begin{bmatrix} 1.5 & -2 & 0.4 \\ 3 & 0.86 & -0.5 \\ 2 & 1.5 & 1.5 \end{bmatrix}.$$

98. $$A = \begin{bmatrix} 0.483030 & 0.510966 & 0.055323 & 0.344944 \\ 0.510966 & -0.617806 & -0.255666 & -0.072740 \\ 0.055323 & -0.255666 & 0.523533 & 0.491080 \\ 0.344944 & -0.072740 & 0.491080 & -0.554933 \end{bmatrix}.$$

99. $$A = \begin{bmatrix} 0.505237 & -0.364336 & 0.079993 & 0.431129 \\ -0.364336 & -0.559546 & -0.490851 & -0.071458 \\ 0.079993 & -0.490851 & 0.472347 & -0.368889 \\ 0.431129 & -0.071458 & -0.368889 & -0.627340 \end{bmatrix}.$$

Using Faddeev's modification of Leverrier's method, determine the eigenvalues and eigenvectors of the following matrices. In addition, use this method to determine the inverses of the matrices 100–103.

100. $$A = \begin{bmatrix} 0.271645 & -0.063420 & 0.090256 & 0.709572 \\ -0.063420 & -0.345194 & -0.722861 & -0.077180 \\ 0.090256 & -0.722861 & 0.360199 & -0.052698 \\ 0.709572 & -0.077180 & -0.052698 & -0.447676 \end{bmatrix}.$$

101. $$A = \begin{bmatrix} -0.078589 & 0.264636 & 0.026244 & 0.770612 \\ 0.264636 & -0.040497 & -0.765021 & -0.069988 \\ 0.026244 & -0.765021 & 0.081177 & 0.264369 \\ 0.770612 & -0.069988 & 0.264369 & -0.134608 \end{bmatrix}.$$

102. $$A = \begin{bmatrix} 0.121330 & 0.090953 & 0.063058 & 0.785000 \\ 0.090953 & -0.214851 & -0.766293 & -0.080056 \\ 0.063058 & -0.766293 & 0.250514 & 0.077005 \\ 0.785000 & -0.080056 & 0.077005 & -0.324027 \end{bmatrix}.$$

103. $$A = \begin{bmatrix} -0.184467 & -0.418863 & 0.074087 & 0.691309 \\ -0.418863 & 0.155609 & -0.690186 & 0.007162 \\ 0.074087 & -0.690186 & -0.084532 & -0.404853 \\ 0.691309 & 0.007162 & -0.404853 & -0.023808 \end{bmatrix}.$$

104. $$D = \begin{bmatrix} 1.111 & 1.222 & 0.333 \\ 1.222 & 1.444 & 0.555 \\ 0.333 & 0.555 & 1.666 \end{bmatrix}, \quad C = \begin{bmatrix} 1 & 1 & 0 \\ 1 & 1 & 0 \\ 0 & 0 & 1 \end{bmatrix},$$

$k = 0(1)\,15.$

105. $$D = \begin{bmatrix} 9.9 & 8.8 & 7.7 & 6.6 \\ 8.8 & 5.5 & 4.4 & 3.3 \\ 7.7 & 4.4 & 2.2 & 1.1 \\ 6.6 & 3.3 & 1.1 & 0.0 \end{bmatrix}, \quad C = \begin{bmatrix} 0.5 & 0 & 0 & 0.5 \\ 0 & 0.5 & 0 & 0 \\ 0 & 0 & 0.5 & 0 \\ 0.5 & 0 & 0 & 0.5 \end{bmatrix},$$

$k = 0(1)\,10.$

106. $$D = \begin{bmatrix} 5.7 & 2.1 & 4.8 & 1.2 \\ 2.1 & 0 & 1.5 & 0.8 \\ 4.8 & 1.5 & 3.6 & 0 \\ 1.2 & 0.8 & 0 & 5.1 \end{bmatrix}, \quad C = \begin{bmatrix} 1 & 0 & 0 & 0 \\ 0 & 1 & 1 & 0 \\ 0 & 1 & 1 & 0 \\ 0 & 0 & 0 & 1 \end{bmatrix},$$

$k = 0(1)\,10.$

107. $D = \begin{bmatrix} 0 & 1.5 & 2 & 4 & 1.8 \\ 1.5 & 3 & -2.3 & 5.1 & 8 \\ 2 & -2.3 & 2 & -8 & -1.2 \\ 4 & 5.1 & -8 & 0 & 4 \\ 1.8 & 8 & -1.2 & 4 & -5 \end{bmatrix}$,

$C = \begin{bmatrix} 0.5 & 1 & 1 & 0 & 0 \\ 1 & 0.5 & 0 & 0 & 0 \\ 1 & 0 & 0.5 & 0 & 0 \\ 0 & 0 & 0 & 0.5 & 0 \\ 0 & 0 & 0 & 0 & 0.5 \end{bmatrix}$,

$k = 0(1)\,10.$

108. $D = \begin{bmatrix} 1 & -2.1 & 0 & -3.2 & 0.8 & -1.9 \\ -2.1 & 0.4 & 1.3 & 4.5 & 2.4 & 1.7 \\ 0 & 1.3 & 2.6 & 1 & 2 & 4.3 \\ -3.2 & 4.5 & 1 & 3.3 & 2.7 & -5.1 \\ 0.8 & 2.4 & 2 & 2.7 & 0.8 & 1.7 \\ -1.9 & 1.7 & 4.3 & -5.1 & 1.7 & 0.9 \end{bmatrix}$,

$C = \begin{bmatrix} 0.3 & 0 & 0 & 0 & 0 & 0.3 \\ 0 & 0.3 & 0 & 0 & 0 & 0 \\ 0 & 0 & 0.3 & 0 & 0 & 0 \\ 0 & 0 & 0 & 0.3 & 0 & 0 \\ 0 & 0 & 0 & 0 & 0.3 & 0 \\ 0.3 & 0 & 0 & 0 & 0 & 0.3 \end{bmatrix}$,

$k = 0(1)\,5.$

109. $D = \begin{bmatrix} 0.23 & 2.11 & 3.72 & 2.03 & -0.98 & 1.23 & 0.75 \\ 2.11 & 0.65 & 1.82 & 0.45 & -3.27 & 0.98 & 1.52 \\ 3.72 & 1.82 & 0.99 & 1.82 & 3.05 & 2.28 & 2.44 \\ 2.03 & 0.45 & 1.82 & 0.65 & 1.51 & 1.72 & -3.91 \\ -0.98 & -3.27 & 3.05 & 1.51 & 0.32 & 1.52 & 0.74 \\ 1.23 & 0.98 & 2.28 & 1.72 & 1.52 & 0.88 & 1.79 \\ 0.75 & 1.52 & 2.44 & -3.91 & 0.74 & 1.79 & 1.11 \end{bmatrix}$,

$C = \begin{bmatrix} 1 & 0 & 0 & 1 & 0 & 0 & 0 \\ 0 & 1 & 0 & 0 & 0 & 0 & 0 \\ 0 & 0 & 1 & 0 & 0 & 0 & 0 \\ 1 & 0 & 0 & 1 & 0 & 0 & 1 \\ 0 & 0 & 0 & 0 & 1 & 0 & 0 \\ 0 & 0 & 0 & 0 & 0 & 1 & 0 \\ 0 & 0 & 0 & 1 & 0 & 0 & 1 \end{bmatrix}$,

$k = 0(1)\,3.$

Determine the characteristic equation for the following matrices using the interpolation method:

110. $$A = \begin{bmatrix} -0.356147 & -0.689921 & 0.066891 & 0.410584 \\ -0.689921 & 0.253173 & -0.406190 & 0.062097 \\ 0.066891 & -0.406190 & -0.250029 & -0.611294 \\ 0.410584 & 0.062097 & -0.611294 & 0.137145 \end{bmatrix}.$$

111. $$A = \begin{bmatrix} -0.031013 & -0.244493 & 0.078546 & 0.762015 \\ -0.244493 & -0.019278 & -0.773625 & -0.033825 \\ 0.078546 & -0.773625 & 0.074933 & -0.241277 \\ 0.762015 & -0.033825 & -0.241277 & -0.194371 \end{bmatrix}.$$

112. $$A = \begin{bmatrix} 0.290063 & -0.793546 & 0.029835 & -0.069869 \\ -0.793546 & -0.382990 & -0.077504 & -0.032665 \\ 0.029835 & -0.077504 & 0.248531 & -0.697272 \\ -0.069869 & -0.032665 & -0.697272 & -0.342839 \end{bmatrix}.$$

113. $$A = \begin{bmatrix} -0.325667 & 0.556907 & -0.038674 & 0.557708 \\ 0.556907 & 0.206584 & -0.557302 & -0.071276 \\ -0.038674 & -0.557302 & -0.163790 & 0.514345 \\ 0.557708 & -0.071276 & 0.514345 & 0.107644 \end{bmatrix}.$$

114. $$D = \begin{bmatrix} 3 & 8 & 5 \\ 8 & 9 & -1 \\ 5 & -1 & 0 \end{bmatrix}, \quad C = \begin{bmatrix} 1 & 1 & 0 \\ 1 & 1 & 0 \\ 0 & 0 & 1 \end{bmatrix},$$

$k = 0(1)5.$

115. $$D = \begin{bmatrix} 9 & 8 & 7 \\ 8 & 6 & 5 \\ 7 & 5 & 4 \end{bmatrix}, \quad C = \begin{bmatrix} 1 & 0 & 1 \\ 0 & 1 & 1 \\ 1 & 1 & 0 \end{bmatrix},$$

$k = 0(0.1)1.5.$

116. $$D = \begin{bmatrix} 0.8 & 5.1 & 2.3 & 7.6 \\ 5.1 & 4.1 & 3.8 & 0 \\ 2.3 & 3.8 & 5.2 & 1.5 \\ 7.6 & 0 & 1.5 & 8.1 \end{bmatrix}, \quad C = \begin{bmatrix} 0.1 & 0 & 0 & 0 \\ 0 & 0.1 & 0.1 & 0 \\ 0 & 0.1 & 0.1 & 0 \\ 0 & 0 & 0 & 0.1 \end{bmatrix},$$

$k = 0(1)5.$

117. $$D = \begin{bmatrix} 3.4 & 5.8 & 9.2 & 5.7 \\ 5.8 & 3.6 & 1.2 & 3.5 \\ 9.2 & 1.2 & 6.7 & 1.4 \\ 5.7 & 3.5 & 1.4 & 0.0 \end{bmatrix}, \quad C = \begin{bmatrix} 1 & 0 & 0 & 1 \\ 0 & 1 & 0 & 0 \\ 0 & 0 & 1 & 0 \\ 1 & 0 & 0 & 1 \end{bmatrix},$$

$k = 0(0.1)1.5.$

118. $$D = \begin{bmatrix} 6.3 & 8.2 & 1.1 & 4.2 & 5.3 \\ 8.2 & 7.9 & 2.4 & 3.3 & 2 \\ 1.1 & 2.4 & 0 & 1.5 & 9.1 \\ 4.2 & 3.3 & 1.5 & 3 & 0 \\ 5.3 & 2 & 9.1 & 0 & 7.1 \end{bmatrix}, \quad C = \begin{bmatrix} 0.2 & 1 & 1 & 0 & 0 \\ 1 & 0.2 & 0 & 0 & 0 \\ 1 & 0 & 0.2 & 0 & 0 \\ 0 & 0 & 0 & 0.2 & 0 \\ 0 & 0 & 0 & 0 & 0.2 \end{bmatrix},$$

$k = 0(1)\,10.$

119. $$D = \begin{bmatrix} 0.32 & -1.44 & 0.45 & 1.67 & 0.76 & 0.83 \\ -1.44 & 0.53 & 1.83 & 1.37 & 0.21 & 0.81 \\ 0.45 & 1.83 & 1.35 & -0.54 & 0.62 & 1.26 \\ 1.67 & 1.37 & -0.54 & 1.04 & 0.91 & -0.79 \\ 0.76 & 0.21 & 0.62 & 0.91 & 1.56 & 0.87 \\ 0.83 & 0.81 & 1.26 & -0.79 & 0.87 & 1.45 \end{bmatrix},$$

$$C = \begin{bmatrix} 0.5 & 0 & 0 & 1 & 0 & 1 \\ 0 & 0.5 & 0 & 0 & 0 & 0 \\ 0 & 0 & 0.5 & 0 & 0 & 0 \\ 1 & 0 & 0 & 0.5 & 0 & 0 \\ 0 & 0 & 0 & 0 & 0.5 & 0 \\ 1 & 0 & 0 & 0 & 0 & 0.5 \end{bmatrix},$$

$k = 0(1)\,5.$

120. $$D = \begin{bmatrix} 7 & 2 & 4 & 6 & 1 & 0 & 3 \\ 2 & 1 & 3 & -8 & 4 & 5 & -4 \\ 4 & 3 & 0 & 5 & -9 & 1 & 4 \\ 6 & -8 & 5 & 7 & 6 & 2 & -5 \\ 1 & 4 & -9 & 6 & 2 & 4 & 1 \\ 0 & 5 & 1 & 2 & 4 & 0 & 2 \\ 3 & -4 & 4 & -5 & 1 & 2 & 8 \end{bmatrix},$$

$$C = \begin{bmatrix} 0.2 & 0 & 0 & 1 & 0 & 0 & 0 \\ 0 & 0.2 & 0 & 0 & 0 & 0 & 0 \\ 0 & 0 & 0.2 & 0 & 0 & 0 & 0 \\ 1 & 0 & 0 & 0.2 & 0 & 0 & 1 \\ 0 & 0 & 0 & 0 & 0.2 & 0 & 0 \\ 0 & 0 & 0 & 0 & 0 & 0.2 & 0 \\ 0 & 0 & 0 & 1 & 0 & 0 & 0.2 \end{bmatrix},$$

$k = 0(1)\,3.$

121. $$A = \begin{bmatrix} 6.214 & 2.180 & 3.184 \\ -1.351 & 8.224 & 5.224 \\ 2.489 & -0.459 & 4.299 \end{bmatrix}.$$

122. $A = \begin{bmatrix} 2 & 2 & 4 & 8 \\ 1 & 4 & 9 & 27 \\ 1 & 5 & 26 & 125 \\ 1 & 4 & 16 & 65 \end{bmatrix}.$

Using **14**, determine the eigenvalues and eigenvectors of the following matrices to six significant places:

123. $A = \begin{bmatrix} -0.168700 & 0.353699 & 0.008540 & 0.733624 \\ 0.353699 & 0.056519 & -0.723182 & -0.076440 \\ 0.008540 & -0.723182 & 0.015938 & 0.342333 \\ 0.733624 & -0.076440 & 0.342333 & -0.045744 \end{bmatrix}.$

124. $A = \begin{bmatrix} 0.522210 & -0.572361 & 0.064652 & 0.177401 \\ -0.572361 & -0.567769 & -0.293492 & -0.067803 \\ 0.064652 & -0.293492 & 0.446261 & -0.549111 \\ 0.177401 & -0.067803 & -0.549111 & -0.596850 \end{bmatrix}.$

125. $A = \begin{bmatrix} 0.480356 & 0.432319 & 0.065252 & 0.429848 \\ 0.432319 & -0.620205 & -0.354667 & -0.080001 \\ 0.065252 & -0.354667 & 0.537108 & 0.424857 \\ 0.429848 & -0.080001 & 0.424857 & -0.571223 \end{bmatrix}.$

Using **14**, determine the eigenvalues and eigenvectors of the following matrices to five significant places:

126. $D = \begin{bmatrix} 0.72 & 0.45 & 0.38 \\ 0.45 & 0.91 & 0.56 \\ 0.38 & 0.56 & 0.12 \end{bmatrix}, \quad C = \begin{bmatrix} 0.1 & 1 & 0 \\ 1 & 0.1 & 0 \\ 0 & 0 & 0.1 \end{bmatrix},$

$k = 0(1)7.$

127. $D = \begin{bmatrix} 1.4 & 1.2 & -1.3 \\ 1.2 & 0.9 & 0.4 \\ -1.3 & 0.4 & 0.8 \end{bmatrix}, \quad C = \begin{bmatrix} 0.2 & 0 & 0.2 \\ 0 & 0.2 & 0.2 \\ 0.2 & 0.2 & 0 \end{bmatrix},$

$k = 0(1)7.$

128. $D = \begin{bmatrix} 0.8832 & -1.5564 & -1.3245 \\ -1.5564 & 0.9911 & -1.2518 \\ -1.3245 & -1.2518 & 0.9738 \end{bmatrix}, \quad C = \begin{bmatrix} 0.5 & 0 & 0.5 \\ 0 & 0.5 & 0.5 \\ 0.5 & 0.5 & 0 \end{bmatrix},$

$k = 0(1)7.$

129. $$D = \begin{bmatrix} -25 & 38 & 15 & 16 \\ 38 & 49 & -18 & 0 \\ 15 & -18 & 19 & 21 \\ 16 & 0 & 21 & -5 \end{bmatrix}, \quad C = \begin{bmatrix} 1 & 0 & 0 & 1 \\ 0 & 1 & 0 & 0 \\ 0 & 0 & 1 & 0 \\ 1 & 0 & 0 & 1 \end{bmatrix},$$

$k = 0(1)5.$

130. $$D = \begin{bmatrix} 1 & -3 & 5 & 7 \\ -3 & 9 & 11 & 13 \\ 5 & 11 & 15 & -17 \\ 7 & 13 & -17 & 19 \end{bmatrix}, C = \begin{bmatrix} 0.5 & 0 & 1 & 1 \\ 0 & 0.5 & 0 & 0 \\ 1 & 0 & 0.5 & 0 \\ 1 & 0 & 0 & 0.5 \end{bmatrix},$$

$k = 0(1)15.$

131. $$D = \begin{bmatrix} 0.32 & 0.87 & 0.63 & 0.38 & 0.51 \\ 0.87 & 0.63 & 0.78 & 0.91 & 0.25 \\ 0.63 & 0.78 & 0.18 & 0.29 & 0.63 \\ 0.38 & 0.91 & 0.29 & 0.72 & 0.46 \\ 0.51 & 0.25 & 0.63 & 0.46 & 0.15 \end{bmatrix},$$

$$C = \begin{bmatrix} 1 & 1 & 1 & 0 & 0 \\ 1 & 1 & 0 & 0 & 0 \\ 1 & 0 & 1 & 0 & 0 \\ 0 & 0 & 0 & 1 & 0 \\ 0 & 0 & 0 & 0 & 1 \end{bmatrix},$$

$k = 0(1)10.$

132. $$D = \begin{bmatrix} 4.11 & 5.32 & 1.97 & -2.88 & 1.44 & 2.65 \\ 5.32 & 3.11 & -4.24 & 3.21 & 2.34 & 1.46 \\ 1.97 & -4.24 & 2.64 & 1.03 & 5.02 & -4.58 \\ -2.88 & 3.21 & 1.03 & 4.33 & 3.72 & 1.26 \\ 1.44 & 2.34 & 5.02 & 3.72 & 4.98 & -5.04 \\ 2.65 & 1.46 & -4.58 & 1.26 & -5.04 & 3.33 \end{bmatrix},$$

$$C = \begin{bmatrix} 1 & 0 & 0 & 1 & 0 & 1 \\ 0 & 1 & 0 & 0 & 0 & 0 \\ 0 & 0 & 1 & 0 & 0 & 0 \\ 1 & 0 & 0 & 1 & 0 & 0 \\ 0 & 0 & 0 & 0 & 1 & 0 \\ 1 & 0 & 0 & 0 & 0 & 1 \end{bmatrix},$$

$k = 0(1)5.$

133. $$D = \begin{bmatrix} 21.1 & -24.4 & -11.3 & 17.4 & 15.3 & 20.1 & 25.2 \\ -24.4 & 15.4 & 16.2 & 14.3 & 10.1 & -19.5 & 18.8 \\ -11.3 & 16.2 & 25.1 & -30.3 & 27.4 & 16.3 & 24.4 \\ 17.4 & 14.3 & -30.3 & 18.2 & 28.1 & 31.5 & 10.8 \\ 15.3 & 10.1 & 27.4 & 28.1 & 13.9 & 14.9 & 11.3 \\ 20.1 & -19.5 & 16.3 & 31.5 & 14.9 & 15.8 & 20.8 \\ 25.2 & 18.8 & 24.4 & 10.8 & 11.3 & 20.8 & 17.8 \end{bmatrix},$$

$$C = \begin{bmatrix} 1 & 0 & 0 & 1 & 0 & 0 & 0 \\ 0 & 1 & 0 & 0 & 0 & 0 & 0 \\ 0 & 0 & 1 & 1 & 0 & 0 & 0 \\ 1 & 0 & 1 & 1 & 1 & 0 & 1 \\ 0 & 0 & 0 & 1 & 1 & 0 & 0 \\ 0 & 0 & 0 & 0 & 0 & 1 & 0 \\ 0 & 0 & 0 & 1 & 0 & 0 & 1 \end{bmatrix},$$

$k = 0(1)3.$

As examples **134–137**, use the power method to determine λ_1 (the maximum modulus eigenvalue) and its corresponding eigenvector to six significant places for the matrices of examples **110–113**. As examples **138–146**, use the power method to determine λ_1 and its corresponding eigenvector to five significant places for the matrices of examples **114–122**.

REFERENCES

[1] Frank Ayres, Jr., *Theory and Problems of Matrices*, Schaum's Outline Series, Schaum Publishing Co., New York, 1962.
[2] G. E. Forsythe and C. B. Moller, *Computer Solution of Linear Algebraic Systems*, Series in Automatic Computation, Prentice-Hall, Englewood Cliffs, N. J., 1967.
[3] C.-E. Fröberg, *Introduction to Numerical Analysis*, Addison-Wesley, Reading, Mass., 1966.
[4] J. H. Wilkinson, *The Algebraic Eigenvalue Problem*, Clarendon Press, Oxford, 1965.
[5] D. K. Faddeev and V. N. Faddeeva, *Computational Methods of Linear Algebra* (English translation of the Russian), W. H. Freeman, San Francisco, 1963.
[6] A. Ralston, *A First Course in Numerical Analysis*, McGraw-Hill, New York, 1965.
[7] I. S. Berezin and N. P. Zhidkov, *Computing Methods* (translation of the Russian), *Volume* 2, Pergamon Press, Oxford, 1965.
[8] V. V. Voevodin, *Numerical Methods in Algebra: Theory and Algorithms* (Russian), Moscow, 1966.

Chapter 3

NUMERICAL SOLUTION OF ORDINARY DIFFERENTIAL EQUATIONS

Since it is often impossible to solve even the simplest ordinary differential equation exactly, approximate methods must be used. By using numerical methods, approximate solutions can be found in tabular form for sequences of values of the argument, where each tabular value is calculated in turn with the help of one or more previous values of the solution and its derivatives. These methods of approximate numerical integration are not exact, and consequently, at each stage of their use, errors are generated which depend on the method, the integration step, length and the properties of the solution.
The best detailed introductory reference to this subject is [1; chapter 6], while the best general advanced reference is [4; chapters 2 and 3]. Additional reading can be obtained from [2; chapter 9], [3; chapter 5], [5; chapters 7 and 8], [6; Part III], [7; chapters 3 and 4], and [8].

1. First order differential equations

THEORY

Consider the differential equation

$$y' = f(x, y) \tag{a}$$

with initial condition

$$y(x_0) = y_0.$$

Given the points $x_1, x_2, \ldots, x_n$ such that $x_0 < x_1 < x_2 \cdots < x_n$, (a) is solved numerically by determining approximate values $y_1, y_2, \ldots, y_n$

for the exact values $y(x_1)$, $y(x_2)$, ..., $y(x_n)$. Different numerical methods for the determination of the values $y_1, y_2, \ldots, y_n$ are discussed below.

In all these methods, neglecting Adams and successive approximation, the *step length* $h_{i+1} = x_{i+1} - x_i$ can be varied for each i. For simplicity of notation, we shall usually write $h_i = h\,(i = 0, 1, 2, \ldots, n-1)$.

1. The Taylor series method {see [5; chapter 8] and [1; § 6.2]}

If $y = y(x)$ is expanded as a Taylor series in the neighbourhood of $x = x_0$, then

$$y(x) \doteq \sum_{i=0}^{m} \frac{y^{(i)}(x_0)}{i!}(x - x_0)^i$$

with error

$$R_m = \frac{y^{(m+1)}(\xi)}{(m+1)!}(x - x_0)^{m+1},$$

$$\xi = x_0 + \vartheta(x - x_0) \qquad (0 < \vartheta < 1).$$

The value m is chosen so that $|R_m|$ is less than some predetermined limit which depends on the required accuracy.

The derivatives $y^{(i)}(x_0)$ are determined by successive differentiation of (a) using the given initial condition:

$$\begin{aligned} y'(x_0) &= f(x_0, y_0), \\ y''(x_0) &= f'_x(x_0, y_0) + f'_y(x_0, y_0)\, y'(x_0), \\ y'''(x_0) &= f''_{x^2}(x_0, y_0) + 2 f''_{xy}(x_0, y_0)\, y'(x_0) + f''_{y^2}(x_0, y_0) y'^2(x_0) \\ &\quad + f'_y(x_0, y_0) y''(x_0). \end{aligned}$$

. . . .

This method can be used to calculate a starting solution for more complex methods – see [7, Note 1].

2. Euler's method {see [1; § 6.3]}

The approximate value y_{k+1} is calculated using

$$y(x_k + h) \approx y_{k+1} = y_k + h f(x_k, y_k),$$

with the error $R_k = |f(x_k) - f_k|$ of order h^2

$$R_k = \frac{h^2}{2} |y''(\xi)| \qquad (x_k \leqq \xi \leqq x_{k+1}).$$

3. The modified Euler method {see [1; § 6.3]}

Using the intermediate values

$$x_{k+\frac{1}{2}} = x_k + \frac{h}{2},$$

$$y_{k+\frac{1}{2}} = y_k + \frac{h}{2} f(x_k, y_k).$$

y_{k+1} is defined by

$$y_{k+1} = y_k + h f\left(x_{k+\frac{1}{2}}, y_{k+\frac{1}{2}}\right).$$

This method is more accurate than Euler's.

4. The modified Euler-Cauchy method

Initially, a first approximation is calculated using Euler's method

$$y_{k+1}^{(0)} = y_k + h f(x_k, y_k).$$

This is then refined using the formula

$$y_{k+1} = y_k + \frac{1}{2} h [f(x_k, y_k) + f(x_{k+1}, y_{k+1})].$$

The error at each step in this method is of order h^3.

5. The modified Euler-Cauchy method with iteration

A first approximation $y_{k+1}^{(0)}$ is calculated using Euler's method

$$y_{k+1}^{(0)} = y_k + h f(x_k, y_k).$$

This is then refined using the iteration

$$y_{k+1}^{(i)} = y_k + \frac{h}{2} [f(x_k, y_k) + f((x_{k+1}, y_{k+1}^{(i-1)})]$$

$$(i = 1, 2, \ldots),$$

until the difference between two successive approximations $y_{k+1}^{(i-1)}$ and $y_{k+1}^{(i)}$ is less than some predetermined accuracy. Then y_{k+1} is

set equal to $y_{k+1}^{(i)}$. However, the difference between $y_{k+1}^{(i)}$ and $y(x_{k+1})$ might not be less than the given accuracy, since the limit of the sequence $y_{k+1}^{(i)}$ $(i = 1, 2, \ldots)$ might not be $y(x_{k+1})$. This is illustrated in some of the problems below.
The error at each step in this method is of order h^3.

6. The Runge-Kutta method {see [1; § 6.5]}

The third order Runge-Kutta method is:

$$y_{k+1} = y_k + \frac{1}{6}(k_1 + 4k_2 + k_3),$$
$$k_1 = hf(x_k, y_k),$$
$$k_2 = hf\left(x_k + \frac{1}{2}h,\ y_k + \frac{1}{2}k_1\right),$$
$$k_3 = hf(x_k + h, y_k - k_1 + 2k_2).$$

The fourth order Runge-Kutta method is:

$$y_{k+1} = y_k + \frac{1}{6}(k_1 + 2k_2 + 2k_3 + k_4),$$
$$k_1 = hf(x_k, y_k),$$
$$k_2 = hf\left(x_k + \frac{1}{2}h, y_k + \frac{1}{2}k_1\right),$$
$$k_3 = hf\left(x_k + \frac{1}{2}h, y_k + \frac{1}{2}k_2\right),$$
$$k_4 = hf(x_k + h, y_k + k_3).$$

When using these methods, the values $k_1, k_2, \ldots$ are determined first.
The Runge-Kutta method of m-th order $(m > 2)$ has errors of order h^{m+1}.

7. Adams' method {see [1; § 6.11]}

The following notation is used

$$f_k = f(x_k, y_k), \quad \Delta^p f_k = \Delta^{p-1} f_k - \Delta^{p-1} f_{k-1}, \quad \Delta^0 f_k = f_k$$
$$(k = 0, 1, 2, \ldots).$$

Adams' extrapolation formula (predictor) is

$$y_{k+1} = y_k + h \sum_{p=0}^{r} \beta_p \, \Delta^p f_k.$$

Its error is

$$R_k = \beta_{r+1} h^{r+2} y^{(r+1)}(\xi) \quad (x_{k-r} \leqq \xi \leqq x_k).$$

The values of the first six coefficients β_p are:

$$\beta_0 = 1,$$
$$\beta_1 = 1/2 = 0.5,$$
$$\beta_2 = 5/12 = 0.416,$$
$$\beta_3 = 3/8 = 0.375,$$
$$\beta_4 = 251/720 = 0.34861,$$
$$\beta_5 = 95/288 = 0.329861.$$

Adams' interpolation formula (corrector) is

$$y^*_{k+1} = y_k + h \sum_{p=0}^{r} \beta^*_p \Delta^p f_{k+1}.$$

Its error is

$$R^*_k = \beta^*_{r+1} h^{r+2} y^{(r+1)}(\eta) \quad (x_{k-r+1} \leqq \eta \leqq x_{k+1}).$$

In order to determine $f_{k+1} = f(x_{k+1}, y_{k+1})$, one uses the value y_{k+1} obtained from the extrapolation formula. The interpolation formula is then used to refine y_{k+1}: y^*_{k+1} is more accurate than y_{k+1}.
The values of the first six coefficients β^*_p are:

$$\beta^*_0 = 1,$$
$$\beta^*_1 = -\frac{1}{2} = -0.5,$$
$$\beta^*_2 = -\frac{1}{12} = -0.083,$$
$$\beta^*_3 = -\frac{1}{24} = -0.0416,$$
$$\beta^*_4 = -\frac{19}{720} = -0.02638,$$
$$\beta^*_5 = -\frac{3}{160} = -0.01875.$$

Note 1. In order to calculate $\Delta^r f_k$, some method for determining the starting values (solution) $y_k, y_{k-1}, \ldots, y_{k-r}$ is needed. The Runge-Kutta method or the method of successive approximations is often used for this purpose.

Note 2. The value of r is chosen so that the differences $\Delta^r f_k$ are constant within the limits of the required accuracy, and hence so, that differences of higher than the rth order can be ignored.
Note 3. If r is fixed, the extrapolation and interpolation formulae reduce to the form

$$y_{k+1} = y_k + h \sum_{p=0}^{r} \alpha_p f_{k-p}$$

$$y^*_{k+1} = y_k + h \sum_{p=0}^{r} \alpha^*_p f_{k+1-p}\,.$$

These are the modified Adams' formulae.
The values of the coefficients α_p and α^*_p for $r = 3$ are:

$$\alpha_0 = \frac{55}{24} = 2.2916, \qquad \alpha^*_0 = \frac{3}{8} = 0.375,$$

$$\alpha_1 = -\frac{59}{24} = -2.4583, \quad \alpha^*_1 = \frac{19}{24} = 0.7916,$$

$$\alpha_2 = \frac{37}{24} = 1.5416, \qquad \alpha^*_2 = -\frac{5}{24} = -0.2083,$$

$$\alpha_3 = -\frac{3}{8} = -0.375, \qquad \alpha^*_3 = \frac{1}{24} = 0.0416.$$

8. The method of successive approximations {see [1; § 6.11]}

This method can be used to construct a starting solution for 7. – see Note 1. above.
It is defined by

$$\begin{aligned} y_1 &= y_0 + h\,\{f_0 + \beta_1 \Delta f_1 + \gamma_2^{(1)} \Delta^2 f_2 + \gamma_3^{(1)} \Delta^3 f_3 + \cdots \\ &\quad + \gamma_r^{(1)} \Delta^r f_r\}, \\ y_2 &= y_1 + h\,\{f_1 + \beta_1 \Delta f_1 + \beta_2 \Delta^2 f_2 + \gamma_3^{(2)} \Delta^3 f_3 + \cdots \\ &\quad + \gamma_r^{(2)} \Delta^r f_r\}, \\ y_3 &= y_2 + h\,\{f_2 + \beta_1 \Delta f_2 + \beta_2 \Delta^2 f_2 + \beta_3 \Delta^3 f_3 + \cdots \\ &\quad + \gamma_r^{(3)} \Delta^r f_r\}, \\ &\cdots\cdots\cdots\cdots\cdots\cdots \\ y_r &= y_{r-1} + h\,\{f_{r-1} + \beta_1 \Delta f_{r-1} + \beta_2 \Delta^2 f_{r-1} + \beta_3 \Delta^3 f_{r-1} \\ &\quad + \cdots + \beta_r \Delta^r f_r\}, \end{aligned}$$

where the β_i are the coefficients of Adams' extrapolation formula and, as in Note 2. above, r is the order at which the differences

$\Delta^r f_k$ are constant within the limits of the required accuracy. The values of the coefficients $\gamma_k^{(i)}$ are:

$$\gamma_2^{(1)} = -\frac{1}{12} = -0.083,$$

$$\gamma_3^{(1)} = \frac{1}{24} = 0.0416,$$

$$\gamma_4^{(1)} = -\frac{19}{720} = -0.02638,$$

$$\gamma_5^{(1)} = \frac{27}{1440} = 0.01875,$$

. . . .

$$\gamma_3^{(2)} = -\frac{1}{24} = -0.0416,$$

$$\gamma_4^{(2)} = \frac{11}{720} = 0.01527,$$

$$\gamma_5^{(2)} = -\frac{11}{1440} = -0.007638,$$

. . . .

$$\gamma_4^{(3)} = -\frac{19}{720} = -0.02638,$$

$$\gamma_5^{(3)} = \frac{11}{1440} = 0.007638,$$

. . . .

$$\gamma_5^{(4)} = -\frac{27}{1440} = -0.01875.$$

. . . .

In order to start this method, it is first necessary to calculate a starting approximation $y_1^{(0)}, y_2^{(0)}, \ldots, y_n^{(0)}$. For example, using the given expression, take the first two terms on the right hand side of the equation for y_1, the first three terms on the right hand side of the equation for y_2, etc. The method can then be used to refine this approximation until two successive approximations $y_j^{(i+1)}$ and $y_j^{(i)}$ $(j = 1, 2, \ldots, n)$ coincide within the limits of the required accuracy. The approximate solution then becomes

$$y_j = y_j^{(i+1)} \qquad (j = 1, 2, \ldots, n).$$

PROBLEMS

1. Solve graphically the following differential equations on the interval [0, 1]. Plot their graphs. In a, d, e and f compare the graphical solution at $x = 1$ with the exact value:

a) $y' = x + y, y(0) = 1$;

b) $y' = x - y^2, y(0) = 1$;

c) $y' = y^2 - 3x^2 - 1, y(0) = 1$;

d) $y' = xy + 1, y(0) = 0$;

e) $y' = -axy, y(0) = 1, a = 0.2\,(0.2)\,1$;

f) $y' = y + a, y(0) = 1, a = 1\,(1)\,5$;

g) $y' = \dfrac{a - xy}{x^2 + y^2 + k}, y(0) = 0, a = 0.6\,(0.1)\,0.9, k = 0.6\,(0.1)\,0.9$.

2. Use the method of Taylor series to solve the differential equation

$$y' = x + y, y(0) = 1,$$

in the neighbourhood of the origin and determine approximate values for the solution at $x = \pm 0.1, \pm 0.2$.

3. Repeat 2. for the differential equations

a) $y' = x - y, y(0) = 0$;

b) $y' = y - \dfrac{2x}{y}, y(0) = 1$;

c) $(1 + x)y' - ay = 0, y(0) = 1, a = 1\,(1)\,5$.

4. Using only the first four terms of the method of Taylor series, solve the following differential equations on the interval [0, 0.2] for the step length $h = 0.05$:

a) $y' = y - x, y(0) = 2$;

b) $y' = -2xy^2, y(0) = 1$.

5. Repeat 4. for the following differential equations on the interval [−0.1, 0] for the step length $h = -0.025$:

a) $y' = xy + 1, y(0) = 0$;

b) $y' = -axy, y(0) = 1, a = 0.2\,(0.2)\,1$.

6. Using the method of Taylor series, solve to five significant places the following differential equations on the interval [0, 0.4] for the step length $h = 0.1$:

a) $y' = 1 - y^2, \quad y(0) = 0;$
b) $y' = 0.5\,y^2 + x^2, \quad y(0) = -1.$

7. Repeat 6. for the following differential equations on the interval $[-0.2, 0]$ for the step length $h = -0.05$:

a) $y' = x + x^2 + y^2, \quad y(0) = 0;$
b) $y' = y + a, \quad y(0) = 1, \quad a = 1\,(1)\,5.$

8. Use Euler's method with step length $h = 0.1$ to solve the following differential equations on the interval $[0, 1]$:

a) $y' = 1 + a\,y \sin x - k\,y^2, \quad y(0) = 0,$
$\quad a = 0.2\,(0.2)\,1, \quad k = 1\,(0.25)\,2;$

b) $y' = \cos(a\,x + y) + k\,(x - y), \quad y(0) = 0,$
$\quad a = 1\,(0.25)\,2, \quad k = 0.5\,(0.25)\,1.5;$

c) $y' = 1 - \sin(a\,x + y) + \dfrac{k\,y}{2 + x}, \quad y(0) = 0,$
$\quad a = 1\,(0.25)\,2, \quad k = -0.3\,(0.2)\,0.5;$

d) $y' = \dfrac{\cos y}{a + x} + k\,y^2, \quad y(0) = 0,$
$\quad a = 1\,(0.25)\,2, \quad k = -0.5\,(0.2)\,0.3;$

e) $y' = -y^2 + \dfrac{a\,x}{1 + x^2}, \quad y(0) = -0.4122,$
$\quad \alpha = 2.5 + \dfrac{a}{40}, \quad a = 14\,(1)\,37.$

9. Use **2**. to solve the following differential equations on the interval $[0, 1]$:

a) 8. c) for step length $h = 0.05$;
b) 8. e) for step length $h = 0.2$.

10. Use **3**. with step length $h = 0.1$ to solve the differential equations in 8. on the interval $[0, 1]$.

11. Using the modified Euler method, solve the following differential equations on the interval $[0, 1]$:

a) 8. c) for step length $h = 0.05$;
b) 8. e) for step length $h = 0.2$.

12. Using **4**. with step length $h = 0.1$, solve the differential equations in 8. on the interval $[0, 1]$.

13. Using the modified Euler-Cauchy method, solve the following differential equations on the interval [0, 1]:

a) 8. c) for step length $h = 0.05$;
b) 8. e) for step length $h = 0.2$.

14. Using 5. with $h = 0.1$, solve to four significant places the differential equations in 8. on the interval [0, 1].

15. Using the modified Euler-Cauchy method with iteration, solve the following differential equations on the interval [0, 1] to four significant places:

a) 8. c) for step length $h = 0.05$;
b) 8. e) for step length $h = 0.2$.

16. Using the third order Runge-Kutta formulae with step length $h = 0.1$, solve the following differential equations on the interval [0, 1]:

a) $y' = \dfrac{a - x + y^2}{1 + ky + x^2}$, $y(0) = 0$,
$a = 0.5\ (0.25)\ 1.5$, $k = 1\ (0.25)\ 2$;

b) $y' = \dfrac{3y^2 + ay + 1}{kxy + 4}$, $y(0) = 0$,
$a = 1\ (0.25)\ 2$, $k = 10\ (2.5)\ 20$;

c) $y' = \dfrac{axy + 1}{ky^2 + 4}$, $y(0) = 0$,
$a = 1\ (0.5)\ 3$, $k = 10\ (2.5)\ 20$;

d) $y' = \dfrac{ay^3 + 1}{ky - x + 5}$, $y(0) = 0$,
$a = 1\ (1)\ 5$, $k = 10\ (2.5)\ 20$.

17. Using the third order Runge-Kutta formulae, solve the following differential equations on the interval [0, 1]:

a) 16. a) for step length $h = 0.05$;
b) 16. d) for step length $h = 0.05$.

18. Using the fourth order Runge-Kutta formulae with $h = 0.1$, solve the following differential equations on the interval [0, 1]:

a) $y' = \dfrac{a - x^2 - y^2}{k + x^2 + xy}$, $y(0) = 0$,
$a = 0.4\ (0.4)\ 2.0$, $k = 1\ (0.25)\ 2$;

b) $y' = \frac{a y^2 + x^2 + 2}{k y + x + 4}$, $y(0) = 0$,

$a = 1\,(0.5)\,3$, $k = 10\,(2.5)\,20$;

c) $y' = \frac{a x y^2 + 1}{k y^2 + x + 5}$, $y(0) = 0$,

$a = 1\,(1)\,5$, $k = 10\,(2.5)\,20$;

d) $y' = \frac{1 - a x^2 - y^2}{k - x y}$, $y(0) = 0$,

$a = 1\,(0.5)\,3$, $k = 1\,(0.5)\,3$.

19. Repeat 18. for the differential equations:

a) 18. b) with $h = 0.05$;

b) 18. d) with $h = 0.05$.

20. Using the fourth order Runge-Kutta formulae, determine to five significant places the value of the approximate solution at $x = 1$ for the following differential equations. The optimal step length h must be chosen on the basis of numerical experimentation with the values $2h$, h, $h/2$ etc.

a) $y' = \frac{a(1 - y^2)}{(1 + k)x^2 + y^2 + 1}$, $y(0) = 0$,

$a = 0.5\,(0.2)\,1.3$, $k = 1\,(0.25)\,2$;

b) $y' = \frac{a + x^2 + y^2}{1 + k x + y^2}$, $y(0) = 0$,

$a = -2\,(0.5)\,0$, $k = 0\,(0.5)\,2$;

c) $y' = a - x + y^2$, $y(0) = 0$,

$a = 0.5\,(0.25)\,1.5$;

d) $y' = -y^2 + \frac{a x}{1 + x^2}$, $y(0) = -0.4122$,

$\alpha = 2.5 + \frac{a}{40}$, $a = 14\,(1)\,38$.

21. Using 8. with $h = 0.05$, determine to five significant places approximate solutions for the following differential equations:

a) $y' = -\frac{a(y + x)^k}{a(y + x)^k + 1}$, $y(0) = 0$,

$a = 1\,(0.3)\,1.6$, $k = 1\,(1)\,4$;

b) $y' = -a x y$, $y(0) = 1$,

$a = 0.2\,(0.2)\,2$;

c) $y' = \frac{a\,x^{a-1}\,y^k}{1 - k\,x^a\,y^{k-1}}, \quad y(2) = 2,$

$a = 2\,(0.2)\,2.8, \quad k = 2.5\,(0.1)\,2.9.$

22. Continue for a further ten steps the numerical integration of the following differential equations using Adams' extrapolation formula:

a) 21. a);
b) 21. c).

23. Continue for a further ten steps the numerical integration of the following differential equations using Adams' interpolation formula:

a) 16. a);
b) 18. b);
c) 18. c);
d) 21. b).

In problems 24. and 25., construct a starting solution using one of the methods **1.**, **6.** or **8.**

24. Use the modified Adams' formulae with $r = 3$ to solve the following differential equations on the interval $[0, d]$ to an accuracy ε. The step length h is chosen so as to yield an approximate solution with the required accuracy ε.

a) $y' = \frac{a\,y + x^2 + 1}{k\,y^2 + x\,y + 4}, \quad y(0) = 0, \quad d = 1, \quad \varepsilon = 10^{-5},$

$a = 1\,(0.5)\,3, \quad k = 10\,(2.5)\,20;$

b) $y' = -\frac{y}{x} - \frac{2\,y^2}{a} \ln x, \quad y(1) = 1, \quad d = 2, \quad \varepsilon = 10^{-4},$

$a = 0.5\,(0.25)\,6.5;$

c) $y' = \frac{1}{a \cos x} - y \operatorname{tg} x, \quad y(0) = 1, \quad d = 1, \quad \varepsilon = 10^{-4},$

$a = 0.2\,(0.01)\,0.44.$

25. Using Adams' method, solve the following differential equations on the interval $[b, d]$ with an accuracy of 10^{-5}:

a) $y' = \frac{x\,y^3}{a} - y, \quad y(0) = 1, \quad b = 0, \quad d = 1,$

$a = 1\,(0.5)\,13;$

b) $y' = \frac{y^2 e^x}{a} - 2y, \quad y(0) = 1, \quad b = 0, \quad d = 1,$

$a = 1\ (0.5)\ 13;$

c) $y' = \frac{a}{x - y^2}, \quad y(1) = 0, \quad b = 1, \quad d = 2,$

$a = -1.1\ (0.1)\ -0.1.$

2. Systems of first order differential equations

THEORY

Consider the system of first order equations {see [5; § 7.3]}

$$y_i' = f_i(x, y_1, y_2, \ldots, y_n) \qquad (i = 1, 2, \ldots, n) \tag{b}$$

and the initial conditions

$$y_1(x_0) = y_{10}, \quad y_2(x_0) = y_{20}, \ldots, y_n(x_0) = y_{n0}. \tag{c}$$

For the numerical solution of (b) and (c), values of the functions $y_1(x)$, $y_2(x)$, ..., $y_n(x)$ are tabulated for various values of the argument x: x_1, x_2, ..., $x_N (x_0 < x_1 < x_2 < \cdots < x_N)$. In this way, n tables of the following form are constructed:

$$\{y_i(x_1), \ldots, y_i(x_N)\} \qquad (i = 1, 2, \ldots, n).$$

All the methods of § 1. can be used to solve (b) and (c), if y_k is replaced by $y_{ik} = y_i(x_k)$ and f_k is replaced by

$$f_{ik} = f_i(x_k, y_{1k}, y_{2k}, \ldots, y_{nk}).$$

For example, Euler's method for first order systems becomes:

$$\{y_{i\,(k+1)} = y_{ik} + h f_{ik}\} \qquad (i = 1, 2, \ldots, n).$$

PROBLEM

26. Solve the following systems of differential equations on the interval [0, 1] with an accuracy ε for the given initial conditions. Since the exact solutions for c, d, e, h, i and k are easily determined, the approximate results can be compared with the exact for these cases.

a) $\begin{cases} y_1' = -x\,y_2, \\ y_2' = \dfrac{y_1}{x}, \end{cases}$

$y_1(0) = 0,\ y_2(0) = 1,\ \varepsilon = 10^{-5}.$

b) $\begin{cases} y_1' = (y_2 - y_1)\,x, \\ y_2' = (y_2 + y_1)\,x, \end{cases}$

$y_1(0) = 1,\ y_2(0) = 1,\ \varepsilon = 10^{-4}.$

c) $\begin{cases} y_1' = y_2 - y_3, \\ y_2' = y_1 + y_2, \\ y_3' = y_1 + y_3, \end{cases}$

$y_1(0) = 1,\ y_2(0) = 2,\ y_3(0) = 3,\ \varepsilon = 10^{-5}.$

d) $\begin{cases} y_1' = y_2 + y_3, \\ y_2' = y_1 + y_3, \\ y_3' = y_1 - y_2, \end{cases}$

$y_1(0) = -1,\ y_2(0) = 1,\ y_2(0) = 0,\ \varepsilon = 10^{-5}.$

e) $\begin{cases} y_1' = a\,y_1 - y_2, \\ y_2' = y_1 + a\,y_2, \end{cases}$

$y_1(0) = 1,\ y_2(0) = 1,\ \varepsilon = 10^{-5},\ a = 0\ (0.1)\ 2.3.$

f) $\begin{cases} y_1' = -y_1 y_2 + \dfrac{\sin x}{x}, \\ y_2' = -y_2^2 + \dfrac{\alpha x}{1 + x^2}, \end{cases}$

$y_1(0) = 0,\ y_2(0) = -0.4122,\ \varepsilon = 10^{-4},$

$\alpha = 2.5 + \dfrac{a}{40},\ a = 25\ (1)\ 48.$

g) $\begin{cases} y_1' = y_2 - (a\,y_1 + k\,y_2)\,y_1, \\ y_2' = e^{y_1} - (y_1 + a\,y_2)\,y_1, \end{cases}$

$y_1(0) = 1,\ y_2(0) = 0,\ \varepsilon = 10^{-5},$

$a = 2\ (0.25)\ 3,\ k = 0.25\ (0.25)\ 1.25.$

h) $\begin{cases} y_1' = y_2 - k\,y_3, \\ y_2' = a\,y_3 - y_1, \\ y_3' = k\,y_1 - a\,y_2, \end{cases}$

$y_1(0) = 1, \quad y_2(0) = 1, \quad y_3(0) = 1, \ \varepsilon = 10^{-4},$

$a = 1\ (0.25)\ 2, \quad k = 1\ (0.25)\ 2.$

i) $\begin{cases} y_1' = 0.25\dfrac{k}{a}(y_2 - y_3), \\ y_2' = 0.75\dfrac{a}{k}(y_3 - y_1), \\ y_3' = a\,k(y_1 - y_2), \end{cases}$

$y_1(0) = 1, \quad y_2(0) = 0, \quad y_3(0) = 0.1, \ \varepsilon = 10^{-5},$

$a = 1\ (0.1)\ 1.4, \quad k = 1.5\ (0.2)\ 2.3.$

j) $\begin{cases} y_1' = \dfrac{k-a}{a} y_2 y_3, \\ y_2' = \dfrac{a+k}{k} y_1 y_3, \\ y_3' = \dfrac{a-k}{a} y_1 y_2, \end{cases}$

$y_1(0) = 1, \quad y_2(0) = 1, \quad y_3(0) = 1, \ \varepsilon = 10^{-4},$

$a = 1\ (0.25)\ 2, \quad k = 2\ (0.25)\ 3.$

k) $\begin{cases} y_1' = a\,y_1 - y_2 + y_3, \\ y_2' = y_1 + k\,y_2 - y_3, \\ y_3' = -\,y_1 + y_2 + a\,y_3, \end{cases}$

$y_1(0) = 1, \quad y_2(0) = 1, \quad y_3(0) = 1, \quad \varepsilon = 10^{-5},$

$a = 1\ (0.3)\ 2.2, \quad k = 1.25\ (0.25)\ 2.25.$

3. Second order differential equations

THEORY

Consider the differential equation {see [3; § 5. 8–1 and bibliographic notes]}

$$y'' = f(x, y, y') \tag{d}$$

and the initial conditions

$$y(x_0) = y_0, \ y'(x_0) = y_0'. \tag{e}$$

To find a numerical solution of this problem, a table of values of $y = y(x)$ must be constructed for some sequence of values of the argument x: $x_1, x_2, \ldots, x_n$ $(x_0 < x_1 < x_2 < \cdots < x_n)$. Sometimes, it is also necessary to tabulate the derivative $y'(x)$.

9. Reduction to first order systems

We introduce the notation

$$y' = z.$$

The solution of (d) and (e) is equivalent to the solution of the following first order system:

$$\begin{cases} y' = z, \\ z' = f(x, y, z), \end{cases}$$

$$y(x_0) = y_0,\ z(x_0) = y_0'.$$

10. The Runge-Kutta method

The fourth order Runge-Kutta method is:

	x	$v_0 = y$	$v_1 = h\,y'$	$k = \frac{h^2}{2} f\left(x, v_0 \frac{v_1}{h}\right)$
1	x_0	v_{00}	v_{10}	k_1
2	$x_0 + \frac{h}{2}$	$v_{00} + \frac{1}{2} v_{10} + \frac{1}{4} k_1$	$v_{10} + k_1$	k_2
3	$x_0 + \frac{h}{2}$	$v_{00} + \frac{1}{2} v_{10} + \frac{1}{4} k_1$	$v_{10} + k_2$	k_3
4	$x_0 + h$	$v_{00} + v_{10} + k_3$	$v_{10} + 2\,k_3$	k_4
6	$x_1 = x_0 + h$	$v_{01} = v_{00} + v_{10} + k^{(0)}$	$v_{11} = v_{10} + k^{(1)}$	
5	$k^{(0)} = \frac{1}{3}(k_1 + k_2 + k_3)$,		$k^{(1)} = \frac{1}{3}(k_1 + 2\,k_2 + 2\,k_3 + k_4)$	

11. Finite difference methods

Faulkner's extrapolation expressions are:

$$y_{k+1} = y_k + h\,y_k' + \frac{1}{2} h^2 \left\{ f_k + \frac{1}{3} \Delta f_k + \frac{1}{4} \Delta^2 f_k + \frac{19}{90} \Delta^3 f_k + \frac{3}{16} \Delta^4 f_k + \frac{863}{5040} \Delta^5 f_k + \cdots \right\},$$

$$y_{k+1}' = y_k' + h \left\{ f_k + \frac{1}{2} \Delta f_k + \frac{5}{12} \Delta^2 f_k + \frac{3}{8} \Delta^3 f_k + \frac{251}{720} \Delta^4 f_k + \frac{95}{288} \Delta^5 f_k + \cdots \right\}.$$

Faulkner's interpolation expressions are:

$$y^*_{k+1} = y_k + h\, y'_k + \frac{h^2}{2}\left\{f_{k+1} - \frac{2}{3}\Delta f_{k+1} - \frac{1}{12}\Delta^2 f_{k+1}\right.$$
$$\left. - \frac{7}{180}\Delta^3 f_{k+1} - \frac{17}{720}\Delta^4 f_{k+1} - \frac{41}{2520}\Delta^5 f_{k+1} - \cdots\right\},$$
$$(y^*_{k+1})' = y_k + h\left\{f_{k+1} - \frac{1}{2}\Delta f_{k+1} - \frac{1}{12}\Delta^2 f_{k+1} - \frac{1}{24}\Delta^3 f_{k+1}\right.$$
$$\left. - \frac{19}{720}\Delta^4 f_{k+1} - \frac{3}{160}\Delta^5 f_{k+1} - \cdots\right\}.$$

The use of these difference schemes follows the same pattern as that for the expressions in Adams' method.

PROBLEM

27. Solve the following second order differential equations with an accuracy of 10^{-5} on the interval $[0, 1]$:

a) $y'' = \cos^2(2\,y) - k\,x^2,\ y(0) = 0,\ y'(0) = a,$
$a = -0.3\ (0.2)\ 0.5,\ k = 0.3\ (0.1)\ 0.7;$

b) $y'' = 1 + (1 - x)\sin^2(k\,y) - \frac{3}{2}x,\ \ y(0) = 0,\ \ y'(0) = a,$
$a = -0.3\ (0.2)\ 0.5,\ k = 1.6\ (0.2)\ 2.4;$

c) $y'' = \cos(3\,x\,y) - k\,x^2,\ y(0) = 0,\ y'(0) = a,$
$a = -0.3\ (0.2)\ 0.5,\ k = 0.7\ (0.1)\ 1.1;$

d) $y'' = 1 - k\sin(3\,x\,y),\ y(0) = 0,\ y'(0) = a,$
$a = -0.3\ (0.2)\ 0.5,\ k = 0.8\ (0.1)\ 1.2;$

e) $y'' = \frac{1}{1 + x} + \frac{1}{2}\sin(k\,y),\ \ y(0) = 0,\ \ y'(0) = a,$
$a = -0.3\ (0.2)\ 0.5,\ k = 0.5\ (0.1)\ 0.9;$

f) $y'' = \frac{x}{1 + x} + \cos(k\,y) - x,\ \ y(0) = 0,\ \ y'(0) = a,$
$a = -0.3\ (0.2)\ 0.5,\ \ k = 0.5\ (0.1)\ 0.9;$

g) $y'' = -a\,y^3,\ \ y(0) = 0.2,\ \ y'(0) = 0,$
$a = 0.2\ (0.1)\ 2.6;$

h) $y'' = -\frac{2}{x}y' - y^a,\ \ y(0) = 1,\ \ y'(0) = 0,$
$a = 1\ (0.5)\ 5.$

4. Boundary value problems for second order differential equations

THEORY

Consider the boundary value problem {see [6; chapter 10] and [3; § 5.8–2 and bibliographic notes]}

$$L[y] = p(x)\,y'' + f(x)\,y' + g(x)\,y = \varphi(x), \tag{f}$$

$$\left.\begin{aligned} \alpha_{10}\,y(a) + \alpha_{11}\,y'(a) + \beta_{10}\,y(b) + \beta_{11}\,y'(b) &= \gamma_1, \\ \alpha_{20}\,y(a) + \alpha_{21}\,y'(a) + \beta_{20}\,y(b) + \beta_{21}\,y'(b) &= \gamma_2. \end{aligned}\right\} \tag{g}$$

The differential equation (f) and the boundary conditions (g) are linear.

12. Reduction to a Cauchy problem

In the particular case when the boundary conditions (g) take the form

$$y(a) = A, \quad y(b) = B,$$

(f) and (g) are equivalent to the two Cauchy problems:

$$\text{I. } L[z] = \varphi(x), \quad z(a) = A, \quad z'(a) = 0,$$
$$\text{II. } L[v] = 0, \quad v(a) = 0, \quad v'(a) = 1.$$

In this way, the solution $y = y(x)$ is obtained as

$$y(x) = z(x) + \frac{B - z(b)}{v(b)}\,v(x).$$

It should be noted that this method sometimes leads to the loss of significant digits.

13. Difference methods

Let the boundary conditions (g) be

$$y(x_0) = y_0, \quad y(x_m) = y_m.$$

Substituting $x = x_i = x_0 + i\,h$ in (f) and replacing the derivatives by the differences

$$y_i' \approx \frac{y_{i+1} - y_i}{h}, \quad y_i'' \approx \frac{y_{i+2} - 2\,y_{i+1} + y_i}{h^2},$$

(f) takes the form

$$y_{i+2} + M_i y_{i+1} + N_i y_i = h^2 \varphi_i$$
$$(i = 0, 1, 2, \ldots, m - 2).$$

An approximate solution is obtained from this expression in two stages:

I. The following recurrence relations are used to calculate the auxiliary quantities c_i and d_i $(i = 0, 1, 2, \ldots, m - 2)$:

$$c_0 = \frac{1}{M_0}, \quad d_0 = h^2 \varphi_0 - N_0 y_0,$$

$$c_{i+1} = \frac{1}{M_{i+1} - N_{i+1} c_i}, \quad d_{i+1} = h^2 \varphi_{i+1} - N_{i+1} c_i d_i.$$

II. The values

$$y_{m-1}, y_{m-2}, \ldots, y_1$$

are calculated using the equation

$$y_{i+1} = c_i (d_i - y_{i+2}).$$

When the boundary conditions (g) are

$$\begin{cases} \alpha_{10} y_0 + \alpha_{11} y_0' = \gamma_1, \\ \beta_{20} y_m + \beta_{21} y_m' = \gamma_2, \end{cases}$$

stages I and II above are replaced by:

$$\text{I.} \quad c_0 = \frac{h \alpha_{10} - \alpha_{11}}{M_0 h \alpha_{10} - \alpha_{11} (M_0 + N_0)},$$

$$d_0 = \frac{h^2 \varphi_0 (h \alpha_{10} - \alpha_{11}) - N_0 h \gamma_1}{h \alpha_{10} - h \alpha_{11}},$$

$$c_{i+1} = \frac{1}{M_{i+1} - N_{i+1} c_i}, \quad d_{i+1} = h^2 \varphi_{i+1} N_{i+1} c_i d_i;$$

$$\text{II.} \quad y_{m-1} = \frac{c_{m-2} (h \beta_{20} + \beta_{21})}{h \beta_{20} + \beta_{21} (1 + c_{m-2})} \left(d_{m-2} - \frac{h \gamma_2}{h \beta_{20} + \beta_{21}}\right),$$

$$y_{i+1} = c_i (d_i - y_{i+2}) \qquad (i = m - 3, m - 4, \ldots, 0).$$

PROBLEM

28. Solve the following boundary value problems with an accuracy of 10^{-5}. In problems b–g, the functions $f_i(x)$ are tabulated for $i = 1\,(1)\,5$:

a) $y'' = -a y - x,$
$y(0) = 0, \; y(1) = 0, \quad a = 0.5\,(0.5)\,6;$

b) $y'' = -f_i(x)\,y' - \cos(a\,x)\,y + 2\,x^2 + 2\,x - 4,$
$y(0) = 0,\ y(1) = 0,\ a = 0.7\ (0.05)\ 0.9.$

Tabulated values of $f_i(x)$

x	$f_1(x)$	$f_2(x)$	$f_3(x)$	$f_4(x)$	$f_5(x)$
−0.2	−1.80365	−1.71833	−1.64072	−1.56982	−1.50479
0.0	−1.79304	−1.70123	−1.61836	−1.54319	−1.47469
0.2	−1.78627	−1.68770	−1.59945	−1.51996	−1.44800
0.4	−1.78322	−1.67763	−1.58385	−1.50000	−1.42458
0.6	−1.78378	−1.67089	−1.57143	−1.48315	−1.40426
0.8	−1.78785	−1.66733	−1.56203	−1.46924	−1.38686
1.0	−1.79530	−1.66683	−1.55551	−1.45814	−1.37223
1.2	−1.80602	−1.66924	−1.55172	−1.44966	−1.36020

c) $y'' = -f_i(x)\,y' - [1 - \sin(a\,x)]\,y - x^2 - x - 1,$
$y(0) = 0,\ y(1) = 0,\ a = 0.2\ (0.1)\ 0.6.$

Tabulated values of $f_i(x)$

x	$f_1(x)$	$f_2(x)$	$f_3(x)$	$f_4(x)$	$f_5(x)$
−0.2	0.97928	0.98052	0.98175	0.98299	0.98423
0.0	1.00000	1.00000	1.00000	1.00000	1.00000
0.2	1.01927	1.01800	1.01674	1.01548	1.01422
0.4	1.03704	1.03448	1.03194	1.02941	1.02689
0.6	1.05328	1.04942	1.04559	1.04179	1.03802
0.8	1.06796	1.06280	1.05769	1.05263	1.04762
1.0	1.08108	1.07463	1.06825	1.06195	1.05572
1.2	1.09264	1.08490	1.07728	1.06977	1.06236

d) $y'' = -f_i(x)\,y' - \cos^2(a\,x)\,y + x^2 + x - 3,$
$y(0) = 0,\ y(1) = 0,\ a = 0.5\ (0.1)\ 0.9.$

Tabulated values of $f_i(x)$

x	$f_1(x)$	$f_2(x)$	$f_3(x)$	$f_4(x)$	$f_5(x)$
−0.2	−1.05197	−1.05544	−1.05894	−1.06246	−1.06600
0.0	−1.01587	−1.01587	−1.01587	−1.01587	−1.01587
0.2	−0.98098	−0.97798	−0.97500	−0.97204	−0.96909
0.4	−0.94718	−0.94159	−0.93607	−0.93061	−0.92522
0.6	−0.91437	−0.90655	−0.89887	−0.89132	−0.88389
0.8	−0.88245	−0.87273	−0.86322	−0.85391	−0.84480
1.0	−0.85135	−0.84000	−0.82895	−0.81818	−0.80769
1.2	−0.82100	−0.80827	−0.79593	−0.78396	−0.77234

e) $y'' = -f_i(x)\,y' - \dfrac{1}{1 + \sin(a\,x)}\,y - x^2 + x - 2,$

$y(0) = 0,\ y(1) = 0,\ a = 1.0\ (0.1)\ 1.4.$

Tabulated values of $f_i(x)$

x	$f_1(x)$	$f_2(x)$	$f_3(x)$	$f_4(x)$	$f_5(x)$
−0.2	0.10949	0.10033	0.09259	0.08596	0.08021
0.0	0.09091	0.08333	0.07692	0.07143	0.06667
0.2	0.07299	0.06689	0.06173	0.05731	0.05348
0.4	0.05535	0.05068	0.04673	0.04335	0.04043
0.6	0.03759	0.03436	0.03165	0.02933	0.02732
0.8	0.01930	0.01761	0.01618	0.01497	0.01393
1.0	0.00000	0.00000	0.00000	0.00000	0.00000
1.2	−0.02092	−0.01894	−0.01730	−0.01592	−0.01475

f) $y'' = -f_i(x)\,y' - \dfrac{1}{1-\sin(ax)}\,y - 2(x^2 + x),$

$y(0) = 0,\ y(1) = 0,\ a = 0.2\,(0.1)\,0.6.$

Tabulated values of $f_i(x)$

x	$f_1(x)$	$f_2(x)$	$f_3(x)$	$f_4(x)$	$f_5(x)$
−0.2	2.13386	2.06607	2.00246	1.94265	1.88631
0.0	2.13115	2.06349	2.00000	1.94030	1.88406
0.2	2.13386	2.06607	2.00246	1.94265	1.88631
0.4	2.14201	2.07384	2.00987	1.94973	1.89309
0.6	2.15567	2.08685	2.02228	1.96158	1.90443
0.8	2.17495	2.10520	2.03978	1.97830	1.92042
1.0	2.20000	2.12903	2.06250	2.00000	1.94118
1.2	2.23103	2.15854	2.09062	2.02685	1.96684

g) $y'' = -f_i(x)\,y' - [1 - \cos(ax)]\,y + x - 2.5,$

$y(0) = 0,\ y(1) = 0,\ a = 0.6(0.1)\,1.0.$

Tabulated values of $f_i(x)$

x	$f_1(x)$	$f_2(x)$	$f_3(x)$	$f_4(x)$	$f_5(x)$
−0.2	0.50063	0.50382	0.50704	0.51031	0.51362
0.0	0.50000	0.50000	0.50000	0.50000	0.50000
0.2	0.50062	0.49755	0.49452	0.49153	0.48857
0.4	0.50243	0.49640	0.49052	0.48478	0.47917
0.6	0.50536	0.49649	0.48792	0.47964	0.47164
0.8	0.50935	0.49772	0.48661	0.47598	0.46581
1.0	0.51429	0.50000	0.48649	0.47368	0.46154
1.2	0.52009	0.50324	0.48745	0.47262	0.45866

h) $y'' = \dfrac{\alpha x}{1 + x^2}\,y + \dfrac{\sin x}{x},$

$y'(0) = -\,0.4122 \cdot y(0),$

$y'(1) = -\,1.1217 \cdot y(1),$

$\alpha = 2.5 + \dfrac{a}{40},\ a = 8\,(1)\,48.$

REFERENCES

[1] B. Carnahan, H. A. Luther and J. O. Wilkes, *Applied Numerical methods*, John Wiley, New York, 1969.

[2] I. S. Berezin and N. P. Zhidkov, *Computing Methods*, Volume 1, Pergamon Press, Oxford, 1965.

[3] A. Ralston, *A First Course in Numerical Analysis*, McGraw Hill, New York, 1965.

[4] L. Collatz, *The Numerical Treatment of Differential Equations*, 3rd Edition, Springer Verlag, Berlin, 1960.

[5] D. G. Moursund and C. S. Duris, *Elementary Theory and Application of Numerical Analysis*, McGraw Hill, New York, 1967.

[6] A. Ralston and H. S. Wilf, *Mathematical Methods for Digital Computers*, Volume 1, John Wiley, New York, 1960.

[7] B. P. Danidovich et. al., *Numerical Methods of Analysis* (Russian), Moscow, 1962.

[8] P. V. Melentlev, *Approximate Calculations* (Russian), Moscow, 1962.

ANSWERS TO THE PROBLEMS OF CHAPTER 1

1.

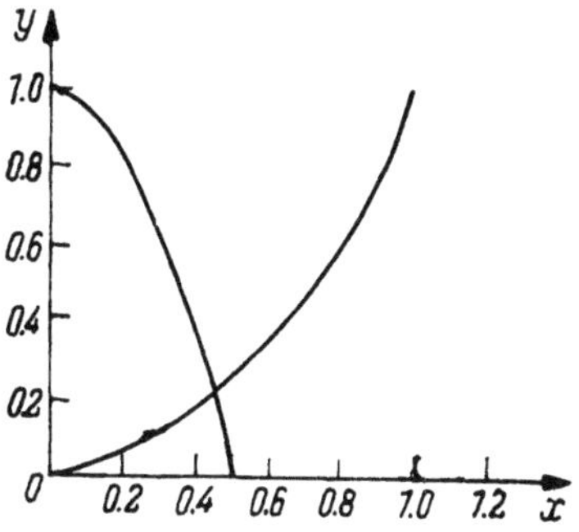

Fig. 1

2.

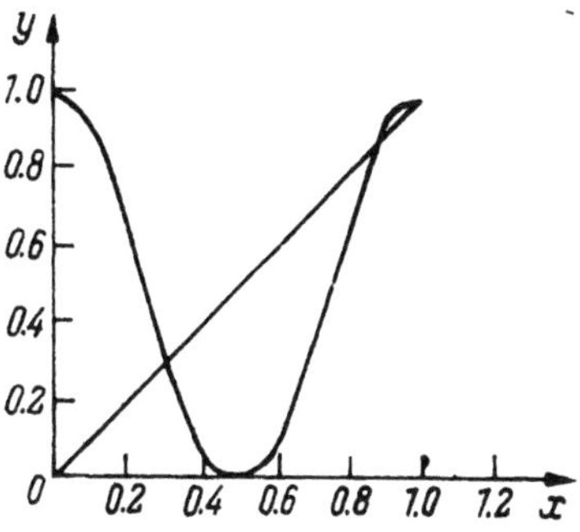

Fig. 2

3.

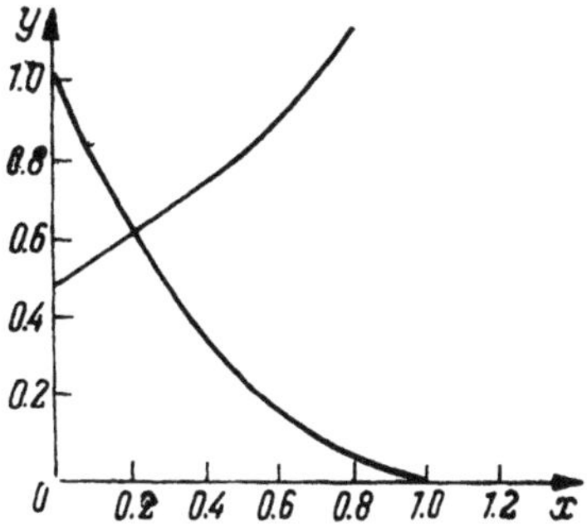

Fig. 3

4.

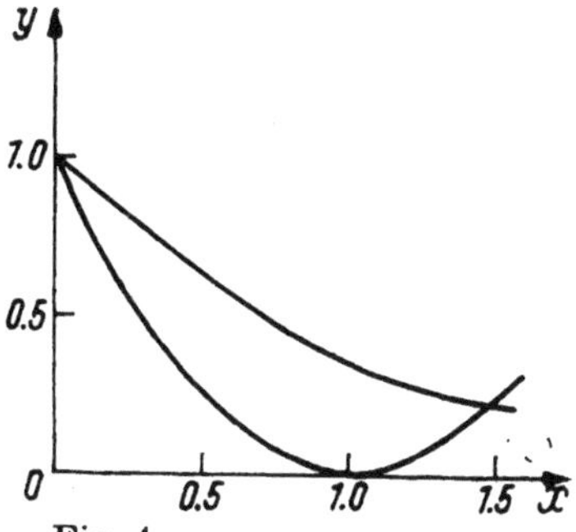

Fig. 4

5.

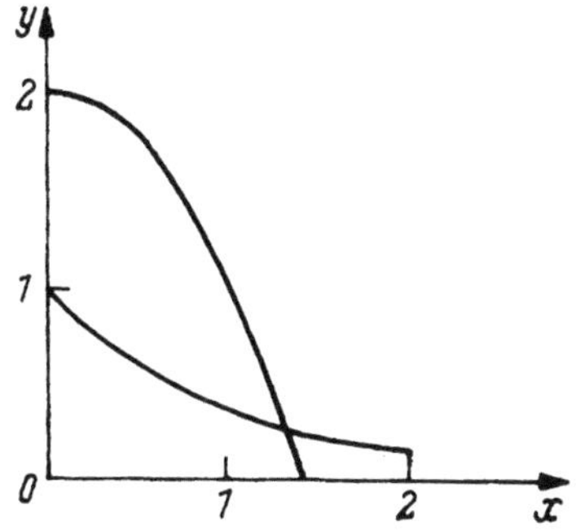

Fig. 5

6.

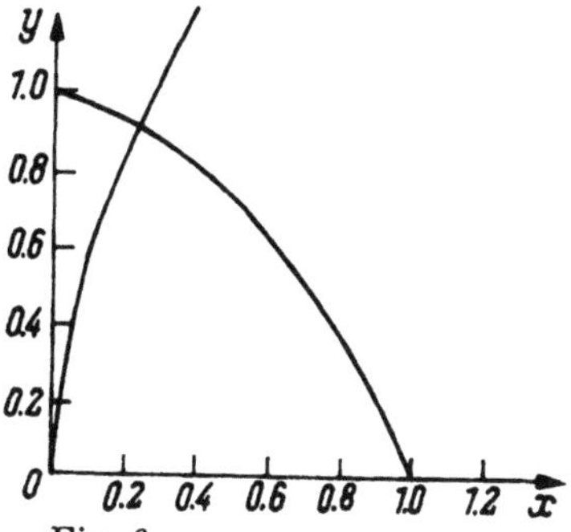

Fig. 6

7.

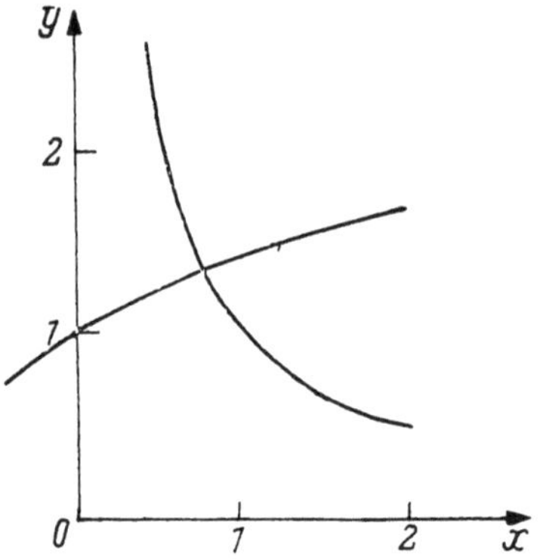

Fig. 7

8.

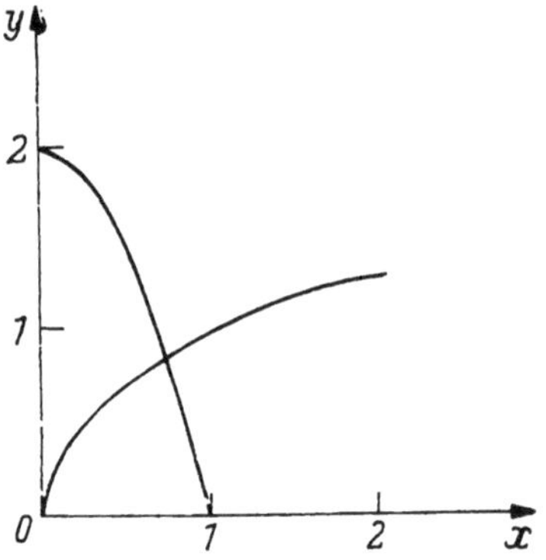

Fig. 8

9.

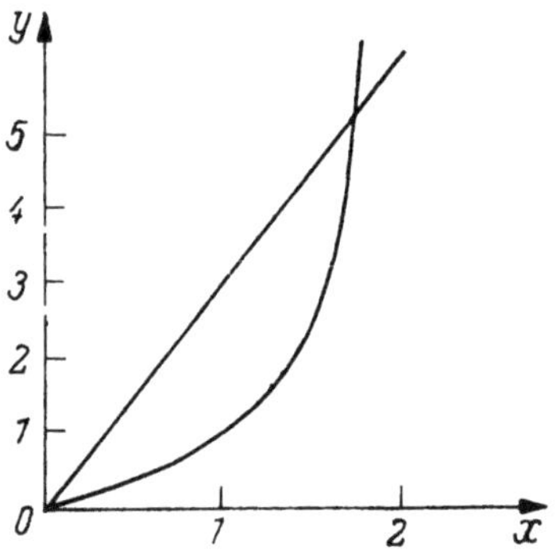

Fig. 9

10.

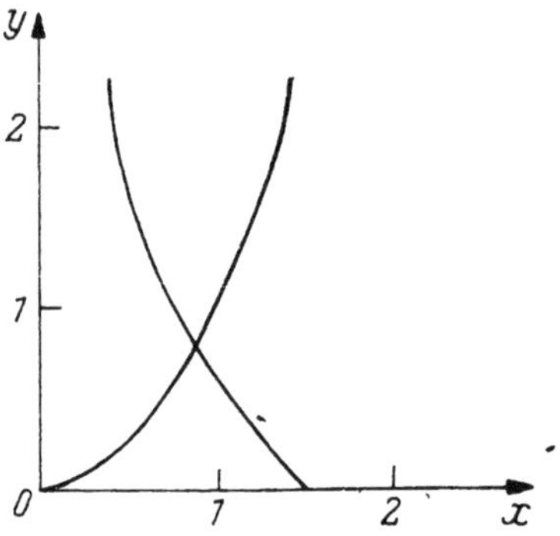

Fig. 10

11.

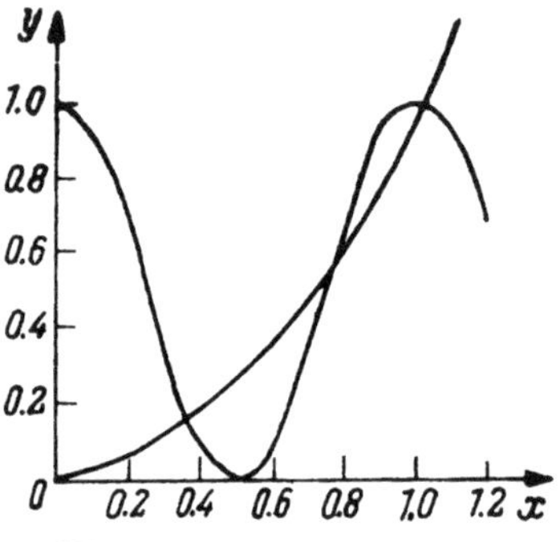

Fig. 11

12.

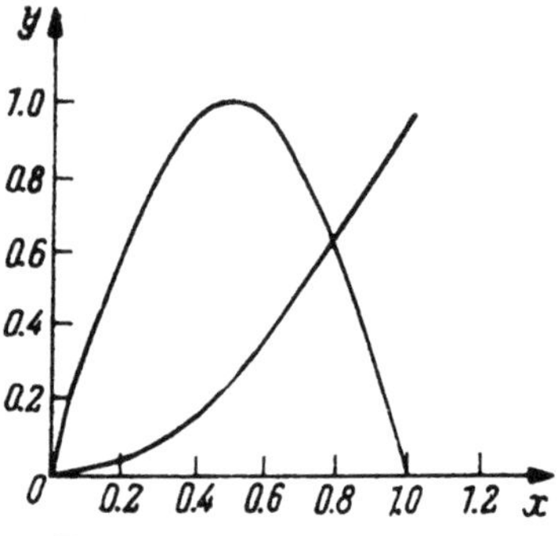

Fig. 12

13. 1.17123. **14.** 0.0; 2.75295. **15.** -0.98737; -0.74656; 0.0; 0.56366. **16.** -0.56669; -0.33443; 0.0; 0.29810. **17.** 0.0; 0.78724. **18.** 0.73909. **19.** -0.43843; 0.43843. **20.** 0.31152; 0.89492; 1.0. **21.** -1.0; -0.78983;

−0.37697; 0.37697; 0.78983; 1.0. **22.** 0.22105. **23.** 0.72088; 3.89780; 4.0. **24.** 0.88678. **25.** 0.93936; 2.72500; 2.99837. **26.** 0.97990. **27.** 0.78112; 2.40135. **28.** 0.0; 0.39928; 6.35234. **29.** 2.92095. **30.** 0.30991; 4.0. **31.** 1.72310. **32.** 1.42153. **33.** 0.39754; 4.68156. **34.** 1.89665. **35.** 3.47301. **36.** 2.74065. **37.** 0.38653. **38.** −0.53727; 1.31597. **39.** −2.51286; 0.0; 1.47767. **40.** −1.31597; 0.53727. **41.** 0.21331. **42.** 0.85261.

43. a	0.5	0.6	0.7	0.8	0.9	1.0
x	0.64119	0.69953	0.76201	0.83078	0.90870	1.0
a	1.1	1.2	1.3	1.4		
x_1	1.11178	1.25773	1.47099	1.88666		
x_2	38.22872	14.76745	7.85706	4.41029		

44. a	1.0	1.1	1.2	1.3	1.4
x_1	0.15859	0.14100	0.12564	0.11216	0.10029
x_2	3.14619	3.29128	3.43361	3.57356	3.71141
a	1.5	1.6	1.7	1.8	1.9
x_1	0.08980	0.08050	0.07224	0.06489	0.05833
x_2	3.84740	3.98171	4.11452	4.24597	4.37618
a	2.0	2.1	2.2	2.3	2.4
x_1	0.05247	0.04723	0.04253	0.03832	0.03455
x_2	4.50524	4.63326	4.76031	4.88647	5.01179
a	2.5	2.6	2.7	2.8	2.9
x_1	0.03115	0.02810	0.02536	0.02289	0.02066
x_2	5.13634	5.26016	5.38330	5.50580	5.62770
a	3.0	3.1	3.2	3.3	3.4
x_1	0.01866	0.01685	0.01523	0.01376	0.01243
x_2	5.74903	5.86982	5.99011	6.10991	6.22926

45. −0.82413. **46.** 1.56519. **47.** 1.57080; 3.90573. **48.** 0.52360; 2.61799. **49.** 1.10251. **50.** 1.45252. **51.** 0.87436. **52.** 4.48175. **53.** 0.87550 **54.** 4.49341. **55.** −5.51881; 1.73499. **56.** 1.76193. **57.** −1.06487. **58.** 2.36502. **59.** 6.16473. **60.** 4.73007. **61.** 1.87510. **62.** 0.73360. **63.** 1.51213. **64.** −0.40103. **65.** 3.78928. **66.** 0.50413.

67. a	1	2	3	4	5
x	0.86033	1.07687	1.19246	1.26459	1.31384
a	6	7	8	9	10
x	1.34955	1.37662	1.39782	1.41487	1.42887
a	11	12	13	14	15
x	1.44058	1.45050	1.45903	1.46643	1.47292
a	16	17	18	19	20
x	1.47864	1.48374	1.48830	1.49241	1.49613
a	21	22	23	24	25
x	1.49951	1.50260	1.50544	1.50804	1.51045

68. a	0.5	0.6	0.7	0.8	0.9
x	0.21995	0.26761	0.30592	0.33672	0.36164
a	1.0		1.1	1.2	1.3
x	0.38197	2.61803	0.39869	0.41256	0.42416
a	1.4	1.5	1.6	1.7	1.8
x	0.43394	0.44222	0.44928	0.45534	0.46055
a	1.9	2.0	2.1	2.2	2.3
x	0.46507	0.46899	0.47241	0.47540	0.47803
a	2.4	2.5	2.6	2.7	2.8
x	0.48034	0.48238	0.48418	0.48578	0.48721

69. n	1	2	3	4
x	0.48587	0.59234	0.66014	0.70676

70. a	1.50	1.55	1.60	1.65	1.70
x	1.44244	1.43219	1.42229	1.41272	1.40347
a	1.75	1.80	1.85	1.90	1.95
x	1.39451	1.38584	1.37742	1.36926	1.36134
a	2.00	2.05	2.10	2.15	2.20
x	1.35364	1.34616	1.33888	1.33180	1.32490

a	2.25	2.30	2.35	2.40	2.45
x	1.31818	1.31164	1.30525	1.29902	1.29294
a	2.50	2.55	2.60	2.65	2.70
x	1.28700	1.28120	1.27554	1.27000	1.26458

71.

a	1.0	1.1	1.2	1.3	1.4
x	2.74371	2.69282	2.63929	2.58291	2.52342
a	1.5	1.6	1.7	1.8	1.9
x	2.46054	2.39394	2.32325	2.24800	2.16767

72. $b = 0.1$

a	0.58	0.56	0.55	0.53	0.52	0.50
x	0.79045	0.76313	0.75019	0.72552	0.71372	0.69105

$b = 0.2$

a	0.56	0.52	0.50	0.47	0.46	0.44
x	0.94551	0.87770	0.84892	0.81007	0.79806	0.77517

$b = 0.3$

a	0.48	0.46	0.44	0.42	0.40	0.37
x	0.98972	0.95480	0.92381	0.89580	0.87014	0.83512

$b = 0.4$

a	0.40	0.38	0.36	0.33	0.31	0.30
x	1.02486	0.99135	0.96144	0.92162	0.89767	0.88635

73. 0.18744. **74.** 0.64119. **75.** 0.75488. **76.** 0.73909. **77.** 0.60710. **78.** 0.22105. **79.** 1.17123. **80.** 0.12243. **81.** 0.67238. **82.** 0.15368; 0.63601. **83.** 29.53659. **84.** 1.55715. **85.** 0.21331. **86.** 0.85261. **87.** 1.25613.

88. 1.32472. **89.** 9.96667. **90.** 1.62936. **91.** 0.78112. **92.** −1.93375. **93.** 0.82413. **94.** 0.93826. **95.** 1.06487. **96.** 2.17157. **97.** 3.78928. **98.** 3.68824. **99.** 0.61906. **100.** 1.31597. **101.** −2.51286; 1.47767. **102.** 1.38233. **103.** 1.06320.

104. 1.95922; 4.55424; 7.94864. **105.** −8.20271; 4.22396; 14.2162. **106.** −4.27161; 8.41684; 15.6545. **107.** −8.41954; −4.14529; 18.1398. **108.** −19.7088; −8.35989; 4.26901. **109.** −22.1517; 4.13927; 8.43749.

110. −24.2165; −4.23999; 8.21913. **111.** −25.6273; −8.76480; −4.07000. **112.** 0.653757; 3.38134; 9.33160. **113.** −7.06744; 0.877473; 3.53193. **114.** −3.68696; 0.660223; 9.52184. **115.** −6.75020; −4.06957; 0.890166. **116.** −9.52684; −0.660014; 4.09175. **117.** −0.889810; 4.55068; 6.66873. **118.** −4.54095; −0.879734; 7.07872. **119.** −9.21794; −4.89361; −0.655153. **120.** 0.655278; 5.11668; 9.19484. **121.** −5.10265; 0.659616; 9.53813. **122.** 2.01853; 4.33124; 6.61233. **123.** −4.33121; 2.01853; 6.81235. **124.** −6.98713; −4.34496; 2.01941.

125. −1.11351; 0.234526; $-0.299163 \pm i \cdot 0.394045$. **126.** −1.88568; −0.516109; $-0.131991 \pm i \cdot 0.843733$. **127.** −2.89701; 0.734620; $-0.482759 \pm i \cdot 1.39467$. **128.** −7.53137; −4.38167; $1.15254 \pm i \cdot 1.81158$. **129.** − 1.09903; −0.677228; $0.635338 \pm i \cdot 2.33044$. **130.** −1.53875; 0.734829; $-1.85606 \pm i \cdot 2.36013$. **131.** −6.89740; −1.73479; $2.29487 \pm i \cdot 2.67039$. **132.** −1.98790; 8.14579; $-3.06818 \pm i \cdot 2.61897$. **133.** 0.467758; 1.15990; $0.0548589 \pm i \cdot 0.246437$. **134.** −0.262635; 0.992493; $-0.370773 \pm i \cdot 0.411450$. **135.** −0.738327; 3.00211; $-1.00794 \pm i \cdot 0.811583$. **136.** 0.992387; 3.87623; $-1.43600 \pm i \cdot 1.48657$. **137.** −4.95774; 10.3107; $-1.96528 \pm i \cdot 1.71153$. **138.** 5.27216; 13.4670; $0.954545 \pm i \cdot 2.74652$. **139.** −8.05868; 1.95755; $-2.17073 \pm i \cdot 3.25278$. **140.** −1.00000; 0.570093; $-0.0503964 \pm i \cdot 0.249024$. **141.** −0.248352; 0.556863; $0.328746 \pm i \cdot 0.817108$. **142.** −3.13531; −0.721828; $-1.23514 \pm i \cdot 1.06738$. **143.** −4.03077; 1.01734; $0.690955 \pm i \cdot 1.50041$. **144.** −9.98106; −4.97132; $-0.649533 \pm i \cdot 2.30371$. **145.** −0.537190; 0.965761; $0.0732815 \pm i \cdot 0.246531$. **146.** 1.75116; 4.52719; $0.366551 \pm i \cdot 1.03949$. **147.** −1.02391; 3.98722; $0.556180 \pm i \cdot 1.80190$. **148.** −4.28026; 10.6920; $-0.347059 \pm i \cdot 3.06932$. **149.** −0.928096; 1.81107; $-0.542683 \pm i \cdot 3.43733$. **150.** 0.996865; 2.01587; $2.19444 \pm i \cdot 3.22593$. **151.** −0.918759; −0.563511; $-0.0686854 \pm i \cdot 0.254374$. **152.** −3.92856; 2.01628; $0.136691 \pm i \cdot 1.07758$. **153.** −0.752467; 0.373157; $0.384507 \pm i \cdot 1.40474$. **154.** −3.85912; −1.04239; $1.42151 \pm i \cdot 1.26776$. **155.** −4.80544; 1.18597; $1.22761 \pm i \cdot 2.26106$. **156.** −10.6215; −6.21638; $-1.13793 \pm i \cdot 3.03115$. **157.** −14.2104; 6.93379; $2.59804 \pm i \cdot 2.58108$. **158.** −1.34642; −1.13489; $1.59677 \pm i \cdot 3.19758$. **159.** −2.05627; 0.983282; $2.41892 \pm i \cdot 2.82607$. **160.** 1.01267; 1.95119; $-0.256681 \pm i \cdot 0.435996$. **161.** 2.61247; 6.89816; $0.533133 \pm i \cdot 1.37932$.

162. −0.385537; 0.764484; $-1.37683 \pm i \cdot 0.655625$. **163.** 0.506786; 1.00934; $-0.801339 \pm i \cdot 1.91465$. **164.** −1.27102; 4.97205; $2.08929 \pm i \cdot 1.57295$. **165.** 1.45711; 6.21310; $2.69565 \pm i \cdot 1.41988$. **166.** −6.81661; 14.2377; $-0.994048 \pm i \cdot 3.20342$. **167.** −0.992405; 2.03388; $-3.22917 \pm i \cdot 2.15773$. **168.** −2.02256; 1.01286; $0.299843 \pm i \cdot 0.393885$. **169.** −4.12687; −1.96302; $0.508310 \pm i \cdot 0.821668$. **170.** −5.92417; 3.12929; $0.409548 \pm i \cdot 1.40194$. **171.** −0.686618; −0.418645; $0.190828 \pm i \cdot 1.45412$. **172.** −1.03261; 0.503198; $-0.738589 \pm i \cdot 1.86753$. **173.** −5.14010; −1.23068; $1.37255 \pm i \cdot 2.36776$. **174.** −6.22891; 1.48178; $1.08333 \pm i \cdot 2.85203$.

175. -15.6538; -5.82242; $2.72321 \pm i \cdot 2.10743$. **176.** -2.13777; -0.963071; $-1.62195 \pm i \cdot 3.53084$. **177.** -0.989324; 2.00214; $-0.106838 \pm i \cdot 0.482732$. **178.** 0.455745; 2.17707; $0.392758 \pm i \cdot 0.765956$. **179.** -3.14170; 6.00377; $-1.10644 \pm i \cdot 0.927340$. **180.** -0.516234; 1.04733; $-1.57709 \pm i \cdot 1.10346$. **181.** 0.643646; 1.23081; $-1.26238 \pm i \cdot 2.26039$. **182.** -1.48623; 6.08116; $1.60714 \pm i \cdot 2.57515$. **183.** 1.62223; 7.63583; $3.03165 \pm i \cdot 1.63398$. **184.** 10.6366; 15.1281; $2.95833 \pm i \cdot 3.31636$. **185.** -6.14128; -2.83744; $1.03247 \pm i \cdot 1.16172$. **186.** -8.24420; 4.00282; $0.830882 \pm i \cdot 1.64967$. **187.** -1.92930; -0.267976; $1.16520 \pm i \cdot 1.57178$. **188.** -1.25402; 0.629820; $1.72727 \pm i \cdot 1.63434$. **189.** -5.78860; -1.52661; $-0.592593 \pm i \cdot 2.80554$. **190.** -6.99706; 1.72783; $2.57792 \pm i \cdot 2.29987$. **191.** -14.7043; -9.32799; $-1.30769 \pm i \cdot 4.07874$. **192.** 0.340868; 0.773560; $-0.285569 \pm i \cdot 0.429780$. **193.** -0.474187; 2.06039; $0.538846 \pm i \cdot 0.792158$. **194.** -4.07288; 8.25145; $0.851064 \pm i \cdot 1.87343$. **195.** -0.624831; 1.25748; $-0.905594 \pm i \cdot 2.21363$. **196.** 0.721062; 1.52894; $-1.41912 \pm i \cdot 2.70907$. **197.** -1.73467; 6.98742; $-0.882716 \pm i \cdot 3.37728$. **198.** 1.94545; 8.22931; $3.30000 \pm i \cdot 1.74452$. **199.** -9.01880; 18.1412; $-1.73809 \pm i \cdot 4.48043$. **200.** 2.55765; 7.73101; $1.57143 \pm i \cdot 4.28809$. **201.** -2.29707; -1.12310; $3.21622 \pm i \cdot 3.08558$. **202.** -2.47115; 10.1283; $3.72727 \pm i \cdot 3.26852$. **203.** -23.9382; 11.6240; $3.15385 \pm i \cdot 4.61346$. **204.** 1.51037; 2.99406; $-4.14286 \pm i \cdot 4.55428$. **205.** -13.8506; 3.55335; $-3.38889 \pm i \cdot 4.87213$. **206.** -15.9097; 4.01498; $3.96154 \pm i \cdot 6.90585$. **207.** -2.24366; 4.50208; $6.95455 \pm i \cdot 5.88509$. **208.** -1.10786; 2.29995; $-2.36207 \pm i \cdot 3.74717$. **209.** -9.53128; 2.57383; $-2.78846 \pm i \cdot 4.00402$. **210.** 14.6775; 19.5983; $-4.34091 \pm i \cdot 3.64594$. **211.** 3.45354; 14.1318; $-1.21429 \pm i \cdot 7.42273$. **212.** -3.73584; -1.60361; $1.28571 \pm i \cdot 6.80786$. **213.** -4.46149; 2.25741; $-7.64286 \pm i \cdot 4.84189$. **214.** 2.43734; 10.0729; $3.19231 \pm i \cdot 3.96639$. **215.** -2.50519; -1.26404; $2.86842 \pm i \cdot 3.85713$. **216.** -27.8640; 14.0068; $5.59091 \pm i \cdot 3.93307$. **217.** -1.70413; 3.48483; $3.69231 \pm i \cdot 6.24358$. **218.** -4.01170; -1.95481; $5.31818 \pm i \cdot 6.06695$. **219.** 1.14482; 2.24880; $3.13636 \pm i \cdot 3.72700$. **220.** -20.5180; -9.53463; $-1.67308 \pm i \cdot 4.76857$. **221.** -1.23171; 2.49362; $-2.57143 \pm i \cdot 3.43749$. **222.** -12.0262; 2.92477; $-3.22222 \pm i \cdot 4.72059$. **223.** -3.48899; 1.71030; $4.26471 \pm i \cdot 5.50299$. **224.** -9.80990; 20.0357; $2.88889 \pm i \cdot 4.04679$. **225.** -2.44448; 1.24448; $3.16129 \pm i \cdot 3.66850$. **226.** 2.82286; 12.4864; $4.04545 \pm i \cdot 4.48206$. **227.** -3.05346; -1.46252; $-3.16667 \pm i \cdot 5.71305$. **228.** 1.60628; 3.68576; $4.02778 \pm i \cdot 5.31761$. **229.** -3.93817; 1.97421; $4.17857 \pm i \cdot 6.46172$. **230.** -2.21588; 9.16460; $4.20000 \pm i \cdot 1.66132$. **231.** -20.1122; 9.85635; $-3.00000 \pm i \cdot 4.15761$. **232.** 1.14694; 2.75629; $3.15714 \pm i \cdot 3.86240$. **233.** -1.49726; 3.03627; $1.63158 \pm i \cdot 5.51754$. **234.** -14.1912; -3.36050; $2.21053 \pm i \cdot 6.85626$. **235.** 1.99525; 3.96737; $-6.64286 \pm i \cdot 4.63075$. **236.** -9.20507; 2.20507; $-2.96875 \pm i \cdot 3.40022$. **237.** 10.4755; 19.2832; $3.57692 \pm i \cdot 3.67605$. **238.** -10.2224; -2.42673; $-3.27778 \pm i \cdot 4.27272$. **239.** -10.6750;

22.8457; $-4.07143 \pm i \cdot 5.05632$. **240.** -3.02991; 1.50106; $-2.15385 \pm i \cdot 5.79584$. **241.** -3.50265; 13.8496; $-3.64286 \pm i \cdot 5.64557$. **242.** -3.96891; 16.0345; $4.81818 \pm i \cdot 6.51318$.

243. ± 1.7321; -0.40000. **244.** -4.0000; 1.0000; 2.0000. **245.** -0.65544; 0.78924; 3.8662. **246.** -2.5321; -1.3473; 0.87939. **247.** 3.2790; $-0.13951 \pm i \cdot 0.94628$. **248.** 1.3247; $-0.66236 \pm i \cdot 0.56228$. **249.** -1.6920; -1.3569; 3.0489. **250.** -3.5043; $1.7522 \pm i \cdot 1.1001$. **251.** -1.2361; 3.2361; $2.0000 \pm i \cdot 1.7321$. **252.** -0.68874; 0.90506; $1.3918 \pm i \cdot 2.4666$. **253.** -2.2340; 0.32762; $0.95320 \pm i \cdot 0.67653$. **254.** $-3.0200 \pm i \cdot 1.0390$; $-0.79999 \pm i \cdot 1.7607$. **255.** -2.9917; -1.0284; 0.044463; 1.9587; 4.0657. **256.** -0.39922; 0.094884; 0.65096; $-0.17331 \pm i \cdot 1.0254$. **257.** -0.16657; $-1.3086 \pm i \cdot 0.62370$; $0.89191 \pm i \cdot 0.79553$. **258.** 0.84431; $-0.40857 \pm i \cdot 0.38702$; $0.98641 \pm i \cdot 2.0033$. **259.** -2.1956; $0.0056769 \pm i \cdot 0.37237$; $1.2271 \pm i \cdot 1.3457$. **260.** -10.898; $-1.3682 \pm i \cdot 1.9092$; $0.31740 \pm i \cdot 0.46397$. **261.** 0.46342; $-0.75293 \pm i \cdot 1.0069$; $0.27122 \pm i \cdot 0.78037$. **262.** -0.94439; $-0.43120 \pm i \cdot 0.86973$; $0.60340 \pm i \cdot 0.87151$. **263.** 1.2663; $-0.71451 \pm i \cdot 1.3076$; $0.58138 \pm i \cdot 1.2001$. **264.** -2.5008; $-0.60304 \pm i \cdot 0.36571$; $0.73343 \pm i \cdot 0.51575$. **265.** -0.56943; $-1.0283 \pm i \cdot 1.4948$; $0.81297 \pm i \cdot 0.49605$. **266.** -7.4415; $-0.40034 \pm i \cdot 0.63300$; $0.62107 \pm i \cdot 0.57701$. **267.** -0.50411; $0.37567 \pm i \cdot 0.88872$; $0.37638 \pm i \cdot 2.0298$. **268.** 1.5035; 2.0784; 2.1201; $1.0990 \pm i \cdot 0.019897$. **269.** -1.0850; 2.2138; $-0.25506 \pm i \cdot 1.6453$; $0.69070 \pm i \cdot 0.75781$. **270.** -0.78691; 6.5198; $-0.13650 \pm i \cdot 0.68495$; $0.77003 \pm i \cdot 0.45412$. **271.** -1.3182; 1.0640; $-0.056091 \pm i \cdot 0.94183$; $0.18320 \pm i \cdot 1.3685$. **272.** 0.38331; 5.2555; $-0.78701 \pm i \cdot 0.57641$; $0.96759 \pm i \cdot 0.32715$. **273.** -1.4305; 1.2963; $-0.11029 \pm i \cdot 1.2669$; $0.17740 \pm i \cdot 0.54953$. **274.** $-2.0051 \pm i \cdot 0.29228$; $-0.18535 \pm i \cdot 0.52054$; $0.078407 \pm i \cdot 1.1359$. **275.** -1.3984; 0.27562; 0.87218; $-0.46566 \pm i \cdot 0.73761$; $0.36297 \pm i \cdot 0.81833$. **276.** -1.9625; 1.1080; 1.5379; $-0.64631 \pm i \cdot 1.1175$; $0.30461 \pm i \cdot 0.99191$.

277. 0.025451; 0.12910; 0.29791; 0.49870; 0.70134; 0.87722; 0.97027. **278.** -0.74447; 0.23714; 0.58002; $0.21943 \pm i \cdot 2.5869$; $0.24423 \pm i \cdot 1.1787$. **279.** -1.7333; 0.66219; 1.2383; $-0.44991 \pm i \cdot 0.42106$; $-0.028709 \pm i \cdot 1.3609$. **280.** -5.1198; 1.0242; $-1.0319 \pm i \cdot 0.54621$; $-0.029962 \pm i \cdot 1.3815$; $0.60965 \pm i \cdot 0.60078$. **281.** ± 0.19509; ± 0.55557; ± 0.83147; ± 0.98079. **282.** -4.9220; -1.0222; 0.60553; 1.7250; $-0.38400 \pm i \cdot 1.1206$; $0.48086 \pm i \cdot 0.72209$. **283.** -2.1833; -0.24009; 2.2664; 4.0168; $0.075641 \pm i \cdot 0.62977$; $0.49447 \pm i \cdot 1.8443$. **284.** $-1.0634 \pm i \cdot 0.38695$; $0.0066105 \pm i \cdot 0.83283$; $0.21827 \pm i \cdot 1.0621$; $0.93937 \pm i \cdot 0.46188$. **285.** -3.1539; -1.6224; -0.73600; $-0.0015921 \pm i \cdot 0.68739$; $0.46322 \pm i \cdot 1.1596$; $0.79456 \pm i \cdot 0.29912$. **286.** -34.081; -0.74918; 1.0471; $-0.66668 \pm i \cdot 1.2592$; $-0.16704 \pm i \cdot 0.83672$; $0.72540 \pm i \cdot 0.55589$. **287.** ± 0.16017; ± 0.27316; ± 1.8015; $\pm (0.26257 \pm i \cdot 0.70126)$. **288.** ± 0.13053; ± 0.38268; ± 0.60876; ± 0.79335; ± 0.92388; $\pm$ 0,99144.

The first row of the answers to problems 289–292 gives the values of x, the second gives the values of y.

289.

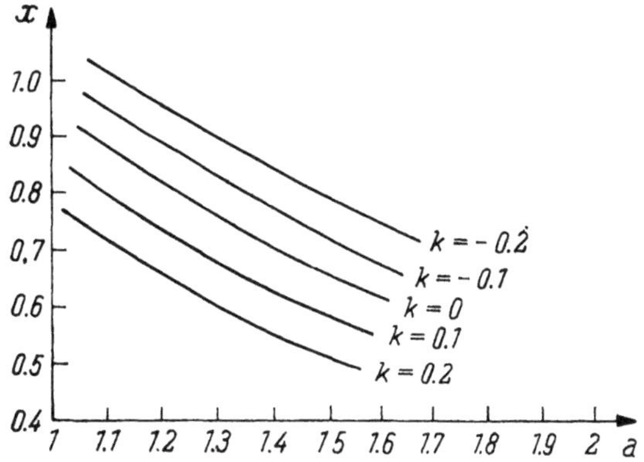

Fig. 13. Graph of x as a function of a for various k

a	k				
	−0.2	−0.1	0.0	0.1	0.2
1.1	0.99386 0.11065	0.95444 0.29842	0.88681 0.46213	0.80470 0.59368	0.71613 0.69796
1.2	0.95683 0.29066	0.89481 0.44645	0.82010 0.57223	0.73960 0.67305	0.65602 0.75474
1.3	0.90164 0.43250	0.83309 0.55314	0.75931 0.65073	0.68280 0.73061	0.60433 0.79673
1.4	0.84421 0.53601	0.77612 0.63059	0.70555 0.70866	0.63332 0.77389	0.55957 0.82878
1.5	0.79063 0.61230	0.72513 0.68861	0.65818 0.75286	0.59000 0.80740	0.52051 0.85385
1.6	0.74217 0.67021	0.67975 0.73344	0.61631 0.78751	0.55183 0.83395	0.48617 0.87386

290.

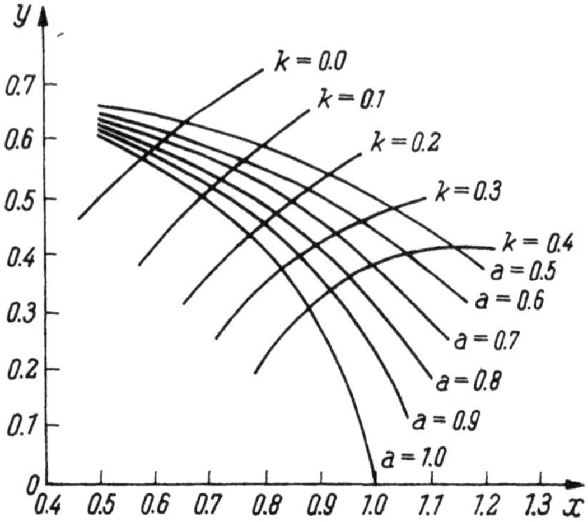

Fig. 14. Family of curves defined by the first and second equations

a	k				
	0.0	0.1	0.2	0.3	0.4
0.5	0.66293 0.62460	0.79656 0.58427	0.91099 0.54085	1,0145 0.49265	1.1077 0.43962
0.6	0.64621 0.61215	0.77208 0.56672	0.87646 0.51918	0.96799 0.46786	1.0484 0.41262
0.7	0.63103 0.60053	0.75057 0.55030	0.84723 0.49877	0.93000 0.44417	1.0013 0.38617
0.8	0.61711 0.58963	0.73135 0.53484	0.82180 0.47943	0.89775 0.42145	0.96195 0.36036
0.9	0.60428 0.57938	0.71396 0.52021	0.79926 0.46102	0.86964 0.39960	0.92814 0.33519
1.0	0.59236 0.56970	0.69807 0.50631	0,77897 0.44340	0.84467 0.37850	0.89836 0.31060

291.

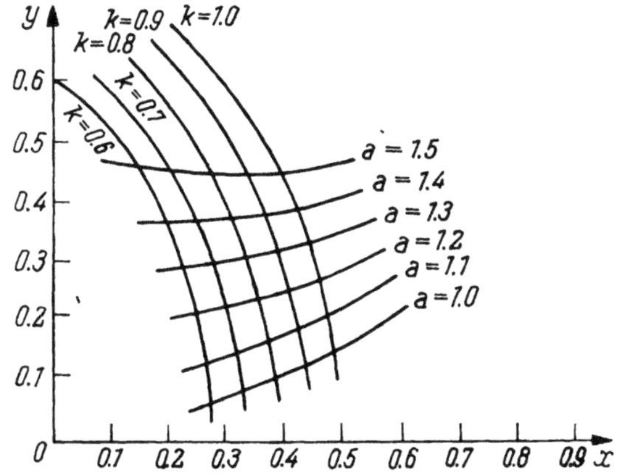

Fig. 15. Family of curves defined by the first and second equations

a	k 0.6	0.7	0.8	0.9	1.0
1.0	0.27241	0.33256	0.38791	0.43933	0.48748
	0.058219	0.082710	0.10778	0.13289	0.15772
1.1	0.26301	0.32249	0.37727	0.42824	0.47606
	0.13345	0.15332	0.17435	0.19588	0.21752
1.2	0.24614	0.30636	0.36161	0.41291	0.46100
	0.20804	0.22311	0.24007	0.25806	0.27655
1.3	0.22065	0.28346	0.34043	0.39298	0.44203
	0.28401	0.29357	0.30607	0.32030	0.33552
1.4	0.18348	0.25202	0.31258	0.36764	0.41856
	0.36450	0.36669	0.37377	0.38367	0.39527
1.5	0.12562	0.20784	0.27566	0.33533	0.38951
	0.45673	0.44606	0.44536	0.44969	0.45692

292.

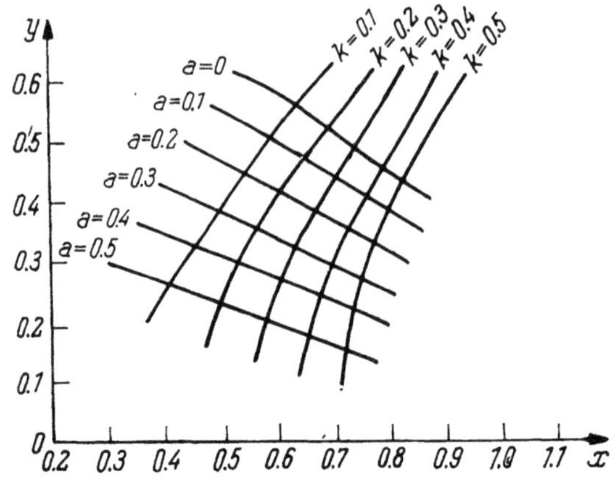

Fig. 16. Family of curves defined by the first and second equations

a	k				
	0.1	0.2	0.3	0.4	0.5
0.0	0.63938 0.55571	0.68926 0.52447	0.73714 0.49333	0.78359 0.46261	0.82901 0.43273
0.1	0.59415 0.50301	0.65030 0.47212	0.70348 0.44146	0.75449 0.41140	0.80389 0.38240
0.2	0.54722 0.44660	0.61075 0.41595	0.66990 0.38571	0.72589 0.35626	0.77952 0.32810
0.3	0.49941 0.38653	0.57167 0.35609	0.63752 0.32623	0.69889 0.29740	0.75694 0.27012
0.4	0.45192 0.32284	0.53443 0.29260	0.60767 0.26317	0.67471 0.23503	0.73728 0.20877
0.5	0.40657 0.25554	0.50088 0.22556	0.58196 0.19666	0.65475 0.16939	0.72169 0.14437

293. $x_1 = 1.4589$, $y_1 = -1.3968$; $x_2 = 3.4874$, $y_2 = 2.2616$. **294.** $x_1 = 0.22684$, $y_1 = 0.36987$; $x_2 = 0.53912$, $y_2 = -0.15049$. **295.** $x_1 = -2.0000$, $y_1 = 2.0000$; $x_2 = 1.3508$, $y_2 = 0.59214$. **296.** $x = 1.7913$, $y = -0.34422$. **297.** $x_1 = 0.75817$, $y_1 = 4.7877$; $x_2 = 1.9549$, $y_2 = -3.0340$. **298.** $x_1 = 0.22875$, $y_1 = 0.088644$; $x_2 = 0.23459$, $y_2 = 0.15797$; $x_3 = 0.31594$, $y_3 = 0.048852$; $x_4 = 0.34203$, $y_4 = 0.26541$; $x_5 = 0.41136$, $y_5 = 0.27125$, $x_6 = 0.45115$, $y_6 = 0.18406$. **299.** $x = 1.0000$, $y = 2.0000$. **300.** $x = 1.0880$, $y = 2.6239$, $z = 2.1427$.

301.							
	a	1.0	1.5	2.0	2.5	3.0	3.5	4.0
	x	1.5020	1.3388	1.2343	1.1590	1.1010	1.0544	1.0156
	y	1.5456	1.6124	1.6615	1.7006	1.7333	1.7613	1.7860
	a	4.5	5.0	5.5	6.0	6.5	7.0	7.5
	x	0.98260	0.95404	0.92893	0.90661	0.88657	0.86842	0.85186
	y	1.8081	1.8280	1.8463	1.8631	1.8787	1.8933	1.9069

ANSWERS TO THE PROBLEMS OF CHAPTER 2

1. 74.1966. **2.** 38.172. **3.** 407.089. **4.** −98,0179. **5.** −23.6223. **6.** 3.83281. **7.** 49.688. **8.** 0.922385. **9.** −24.3430. **10.** 0.554472. **11.** 139036. **12.** −4077.07. **13.** 1219.05. **14.** −17.0314. **15.** 1.04738. **16.** 8782.56. **17.** 2976.29. **18.** 0.194999. **19.** 136.903. **20.** −10.9974. **21.** 856.518. **22.** −4.92843. **23.** 8286.93. **24.** 1047.62.

25. $$\begin{bmatrix} 0.6 & -0.4 & 0.2 & -0.2 \\ 0.4 & 0.4 & -0.2 & 0.2 \\ 0.2 & 0.2 & 0.4 & -0.4 \\ 0.2 & 0.2 & 0.4 & 0.6 \end{bmatrix}.$$

26. $$\begin{bmatrix} -0.223543 & -0.527690 & 0.086324 & -0.716677 & -0.987278 \\ -0.653185 & -0.349254 & -0.011576 & -0.022467 & -0.648328 \\ -0.444679 & -0.237767 & -0.392505 & -0.070676 & -0.023127 \\ -0.346294 & -0.185162 & -0.305664 & -0.645002 & -0.303975 \\ -0.198099 & -0.105922 & -0.174856 & -0.368975 & -0.610571 \end{bmatrix}.$$

27. $$\begin{bmatrix} -0.24921 & 0.27657 & -0.33976 & 0.25471 & 0.35195 & -1.2315 \\ 0.38512 & -0.50061 & 0.33272 & -0.16919 & -0.58366 & 1.9132 \\ 0.77175 & -0.71827 & 0.63317 & -0.19286 & -1.2913 & 3.7649 \\ -1.8209 & 2.0916 & -1.7011 & 0.81147 & 3.3887 & -9.6969 \\ 1.6196 & -1.9212 & 1.4940 & -0.61599 & -3.2292 & 9.1724 \\ -0.32489 & 0.37631 & -0.48838 & 0.27399 & 0.51121 & -1.5568 \end{bmatrix}.$$

28.

0.041838	−0.12783	−0.0066557	0.0087649	0.10078	−0.075768	0.079572
0.19845	0.066359	−0.18662	−0.20837	0.10367	0.31103	−0.22128
−0.33056	0.013287	0.21115	0.28821	−0.067847	−0.020256	−0.21268
−0.015155	0.086811	−0.11922	0.014322	−0.040830	0.097228	0.023791
0.067775	−0.0023626	0.035322	0.063321	−0.057892	−0.23770	0.18186
0.085761	−0.011327	0.010458	−0.063697	−0.095234	0.10447	0.028806
−0.012541	−0.030290	0.089572	−0.10432	0.086239	−0.15329	0.14639

29.

1.5769	0.98196	−0.89835	−0.95872	−1.4154	−0.51029	0.31490	2.6567
2.4191	0.088553	−0.85777	−2.3774	1.9534	−1.7403	−0.49128	0.86731
−1.7157	0.062715	1.5219	2.2373	−1.0316	1.0136	0.33283	−1.8135
−1.7066	−0.20772	0.76492	2.1057	−0.31633	1.0630	0.22407	−2.1174
0.71431	−0.49252	−0.55856	−0.62733	0.46273	−0.99894	0.015780	0.97092
1.7829	−0.83765	−0.87685	−1.7123	1.4928	−1.3554	−0.27386	1.1421
−1.6010	0.29221	0.49110	0.81131	−0.93989	1.5820	0.49802	−0.16198
−1.4159	−0.24158	0.56871	0.89342	0.27554	1.1568	−0.53054	−0.84638

30.

0.033551	−0.047736	0.077643	−0.16822	−0.020044	0.10840	−0.050685	−0.040052	0.17124
0.17124	0.033551	−0.047736	0.077643	−0.16822	−0.020044	0.10840	−0.050685	−0.040052
−0.040052	0.17124	0.033551	−0.047736	0.077643	−0.16822	−0.020044	0.10840	−0.050685
−0.050685	−0.040052	0.17124	0.033551	−0.047736	0.077643	−0.16822	−0.020044	0.10840
0.10840	−0.050685	−0.040052	0.17124	0.033551	−0.047736	0.077643	−0.16822	−0.020044
−0.020044	0.10840	−0.050685	−0.040052	0.17124	0.033551	−0.047736	0.077643	−0.16822
−0.16822	−0.020044	0.10840	−0.050685	−0.040052	0.17124	0.033551	−0.047736	0.077643
0.077643	−0.16822	−0.020044	0.10840	−0.050685	−0.040052	0.17124	0.033551	−0.047736
−0.047736	0.077643	−0.16822	−0.020044	0.10840	−0.050685	−0.040052	0.17124	0.033551

31.
$$\begin{bmatrix} 0.43948 & -0.22818 & 0.57062 & 0.32919 \\ 0.32919 & 0.43948 & -0.22818 & 0.57062 \\ 0.57062 & 0.32919 & 0.43948 & -0.22818 \\ -0.22818 & 0.57062 & 0.32919 & 0.43948 \end{bmatrix}.$$

32.
$$\begin{bmatrix} 0.073715 & 0.13094 & 0.038409 & -0.029486 & -0.070029 \\ 0.022308 & 0.36857 & 0.17478 & 0.39108 & -0.071193 \\ 0.021339 & 0.22211 & 0.15849 & 0.19146 & 0.079720 \\ 0.027158 & 0.10087 & 0.25626 & 0.38914 & 0.17420 \\ 0.10572 & -0.035887 & -0.023860 & -0.042289 & 0.14956 \end{bmatrix}.$$

33.
$$\begin{bmatrix} -0.37047 & 2.2939 & -2.6274 & 1.0131 & -5.0878 & 4.8380 \\ 1.2217 & -3.7848 & 4.0474 & -1.4439 & 8.0013 & -7.8535 \\ -0.66941 & 2.0531 & -1.7406 & 0.51068 & -3.7474 & 3.6325 \\ 0.23625 & -1.1669 & 1.2576 & -0.30882 & 2.0074 & -1.8995 \\ 0.38650 & -0.48003 & 0.21367 & 0.37872 & 1.2221 & -1.5953 \\ -0.70151 & 1.5202 & -1.4675 & 0.17050 & -3.0808 & 3.5863 \end{bmatrix}.$$

$$\begin{bmatrix} 1.1366 & 0.99573 & 2.2164 & -2.0591 & -3.7196 & 2.3816 & -0.32917 \\ 2.9709 & -3.6493 & 0.54563 & -4.5589 & 2.3383 & 2.7806 & 0.11935 \\ -6.2796 & 6.0604 & -6.8264 & 8.6894 & 2.3198 & -6.9618 & 0.60215 \\ 3.2544 & -0.21226 & 4.1538 & -3.1413 & -4.2801 & 1.9162 & -0.26491 \\ 4.7273 & -2.5549 & 5.7633 & -10.177 & -2.0185 & 4.4424 & -0.29303 \\ -5.9686 & 1.2958 & -7.6066 & 10.209 & 5.1258 & -4.9985 & 1.8523 \\ -1.6610 & -0.88058 & -0.23989 & 4.5696 & 1.8692 & -0.83925 & -1.2645 \end{bmatrix}.$$

35. $$\begin{bmatrix} 1.3651 & 2.3397 & -0.72887 & -0.52296 & -0.25908 & -0.054438 & -0.41567 & -1.4064 \\ -0.16351 & -0.40962 & 0.082057 & 0.16239 & 0.19846 & 0.18305 & 0.11347 & -0.084542 \\ -2.4228 & -3.6034 & 1.3221 & -0.21516 & 0.18434 & 0.69655 & 1.9599 & 3.2898 \\ 0.10728 & -0.48666 & -0.015884 & 1.9846 & 0.042864 & 0.093998 & -0.18871 & -1.5911 \\ 2.3994 & 7.6334 & -1.2453 & -2.2163 & -0.60990 & -2.3878 & -2.3538 & -2.4194 \\ -0.59746 & -2.6766 & 0.26145 & 0.36912 & 0.21598 & 1.8353 & -0.21218 & 1.1032 \\ -0.77366 & -2.5493 & 0.36248 & 0.44360 & 0.20820 & -0.26327 & 1.7385 & 1.2203 \\ 0.29965 & 1.1117 & -0.12401 & -0.15077 & -0.090362 & -0.058578 & -0.56832 & -0.56909 \end{bmatrix}.$$

36. $$\begin{bmatrix} 2.4239 & 1.6543 & -3.1317 & 2.6836 & -5.1349 & 1.7059 & -0.049964 & -0.082097 & 1.8368 \\ 2.3195 & 1.7300 & -2.8963 & 2.6121 & -4.5871 & 2.0761 & -0.22444 & -0.72629 & 1.7454 \\ 2.0934 & 1.3763 & -2.5873 & 1.8624 & -3.8932 & 1.7121 & -0.69395 & -0.88134 & 2.1189 \\ 1.9664 & 1.6512 & -3.1070 & 2.7423 & -4.5180 & 1.5333 & -0.17446 & -0.14290 & 1.7521 \\ -3.3881 & -2.6591 & 5.4217 & -4.9234 & 8.9185 & -2.9552 & -0.12823 & -0.41304 & -2.9049 \\ 1.7954 & 1.4962 & -2.9783 & 2.1722 & -4.0375 & 1.6643 & -0.24143 & -0.16635 & 1.7323 \\ 1.4459 & 0.54936 & -1.7776 & 1.7618 & -2.7439 & 1.0369 & 0.32198 & -0.036294 & 0.82182 \\ -1.0115 & -0.70093 & 1.3100 & -0.68381 & 1.6151 & -0.84519 & 0.33591 & 0.94266 & -1.1341 \\ -2.7698 & -1.6215 & 3.3322 & -2.8640 & 4.7189 & -1.8789 & 0.42777 & 0.69920 & -1.5527 \end{bmatrix}.$$

37. $$\begin{bmatrix} 0.16332 & 1.5100 & -1.3372 & -0.37605 & -0.83017 & -0.22212 & -0.078988 & 0.20425 & 0.22851 & 0.069446 \\ 0.050567 & -1.7364 & 1.3721 & 0.60612 & 1.0002 & 0.61088 & 0.50137 & -0.19493 & -0.47749 & -0.049772 \\ -0.53405 & 2.8405 & -1.8200 & -0.50141 & -1.2124 & -0.79191 & -0.45051 & 0.43758 & 0.59373 & -0.098524 \\ 0.46992 & -4.2482 & 3.6435 & 1.0976 & 2.0892 & 1.3891 & 0.96442 & -0.69679 & -1.1375 & 0.088962 \\ 0.20426 & -0.82904 & 0.96642 & -0.025404 & 0.54003 & 0.53054 & 0.56701 & -0.11179 & -0.46165 & 0.092442 \\ 0.44590 & -4.5187 & 3.6877 & 1.3180 & 2.5429 & 1.3025 & 0.60602 & -0.86031 & -0.99790 & 0.077080 \\ 0.17325 & -1.5663 & 1.4232 & 0.43891 & 0.99571 & 0.25106 & 0.33797 & -0.22486 & -0.38850 & 0.13592 \\ -0.22437 & 1.3612 & -1.0527 & -0.33279 & -0.70762 & -0.39531 & -0.27725 & 0.42831 & 0.17741 & -0.029099 \\ 0.42669 & -3.6116 & 2.8308 & 1.0113 & 1.8832 & 0.96484 & 0.65581 & -0.38698 & -0.75788 & -0.19842 \\ -0.21254 & 2.5560 & -2.7415 & -0.85566 & -1.7716 & -0.90604 & -0.86666 & 0.62623 & 1.2137 & 0.18613 \end{bmatrix}.$$

38.

k	x_1	x_2	x_3	$\|D + kC\|$
0	1.34023	−4.75801	2.57775	80.37
1	1.42684	−4.68990	2.58680	80.762
2	1.51455	−4.61325	2.58654	81.302
3	1.60275	−4.52830	2.57698	81.996
4	1.69081	−4.43543	2.55821	82.85
5	1.77810	−4.33508	2.53045	83.87
6	1.86401	−4.22782	2.49400	85.062
7	1.94795	−4.11429	2.44928	86.432
8	2.02935	−3.99520	2.39677	87.986
9	2.10773	−3.87131	2.33704	89.73
10	2.18260	−3.74343	2.27074	91.67
11	2.25358	−3.61236	2.19855	93.812
12	2.32034	−3.47894	2.12117	96.162
13	2.38261	−3.34396	2.03935	98.726
14	2.44020	−3.20822	1.95383	101.51
15	2.49296	−3.07245	1.86533	104.52

39.

k	x_1	x_2	x_3	x_4	$\|D + kC\|$
0	2.6125	−3.1375	3.0250	−1.6875	−320.00
1	3.5389	−3.9474	2.9519	−1.4246	−247.97
2	5.2475	−3.6020	2.3872	−0.74870	−195.67
3	7.3825	−1.3106	1.0741	0.44881	−171.60
4	8.7969	2.3239	−0.58590	1.7469	−179.72
5	8.8991	5.3778	−1.7812	2.5573	−219.37
6	8.2840	7.0726	−2.3164	2.8555	−285.36
7	7.6089	7.8560	−2.4603	2.9145	−367.88
8	7.1618	8.2907	−2.4570	2.9381	−452.56
9	7.0575	8.7638	−2.4414	3.0437	−520.45
10	7.4416	9.6241	−2.5000	3.3431	−548.00
11	8.7539	11.531	−2.7590	4.0823	−507.11
12	12.983	17.089	−3.6895	6.2868	−365.09
13	58.620	75.649	−14.204	29.575	−84.666
14	−13.543	−16.761	2.5393	−7.1620	376.01
15	−4.8057	−5.5369	0.55742	−2.6886	1063.4

40.

k	x_1	x_2	x_3	x_4	x_5	$\|D+kC\|$
0	1.0000	1.0000	1.0000	1.0000	1.0000	83.323
1	0.93730	0.95044	0.84892	0.87570	0.91594	109.75
2	0.89156	0.90956	0.72285	0.76792	0.85028	140.18
3	0.85802	0.87623	0.61453	0.67301	0.79882	174.41
4	0.83382	0.84955	0.51905	0.58810	0.75857	212.14
5	0.81717	0.82881	0.43292	0.51093	0.72744	252.93
6	0.80697	0.81349	0.35352	0.43964	0.70393	296.23
7	0.80257	0.80324	0.27877	0.37266	0.68700	341.38
8	0.80362	0.79790	0.20695	0.30866	0.67597	387.58
9	0.81007	0.79744	0.13650	0.24637	0.67042	433.94
10	0.82208	0.80202	0.065956	0.18461	0.67021	479.44
11	0.84011	0.81200	−0.0062096	0.12215	0.67540	522.94
12	0.86487	0.82795	−0.081667	0.057613	0.68634	563.17
13	0.89747	0.85079	−0.16239	−0.010570	0.70365	598.77
14	0.93953	0.88183	−0.25083	−0.084368	0.72837	628.24
15	0.99340	0.92302	−0.35027	−0.16636	0.76207	619.97

41.

k	x_1	x_2	x_3	x_4	x_5	x_6
0.0	1.0000	0.0000	−1.0000	0.0000	1.0000	0.0000
0.2	0.91696	0.25088	−1.1237	0.027792	1.0873	−0.23222
0.4	0.86111	0.45208	−1.2515	0.056985	1.1344	−0.39149
0.6	0.81982	0.63403	−1.3995	0.094418	1.1674	−0.51968
0.8	0.78745	0.81637	−1.5832	0.14494	1.2010	−0.64020
1.0	0.76130	1.0169	−1.8225	0.21390	1.2472	−0.77027
1.2	0.74010	1.2573	−2.1477	0.30905	1.3187	−0.92714
1.4	0.72343	1.5699	−2.6093	0.44314	1.4343	−1.1333
1.6	0.71116	2.0095	−3.2970	0.63810	1.6256	−1.4232
1.8	0.70257	2.6722	−4.3720	0.93043	1.9511	−1.8507
2.0	0.69170	3.6925	−6.0633	1.3586	2.5037	−2.4665
2.2	0.64884	4.8940	−8.0821	1.7764	3.2413	−3.0329
2.4	0.52184	4.3583	−7.1904	1.2852	3.1237	−2.2106
2.6	0.41653	1.6989	−2.6873	0.00073363	1.6875	−0.26965
2.8	0.39888	0.17283	−0.077105	−0.60946	0.74644	0.61944
3.0	0.40349	−0.32463	0.79180	−0.75394	0.37959	0.81965

k	$\|D + kC\|$	k	$\|D + kC\|$
0.0	0.36597	2.6	0.15251
0.2	0.43182	2.8	0.31250
0.4	0.48004	3.0	0.61673
0.6	0.50812		
0.8	0.51409		
1.0	0.49698		
1.2	0.45731		
1.4	0.39755		
1.6	0.32274		
1.8	0.24102		
2.0	0.16427		
2.2	0.10876		
2.4	0.095804		

42. $x_1 = 0.18691$, $x_2 = 0.60228$, $x_3 = 0.52468$, $x_4 = -0.10430$, $x_5 = 0.35417$, $x_6 = 0.65180$, $x_7 = 1.1075$; $|A| = 0.14912$. **43.** $x_1 = 0.74969$, $x_2 = 1.0958$, $x_3 = 1.1209$, $x_4 = 0.97042$, $x_5 = 0.09922$, $x_6 = 1.0959$, $x_7 = 0.75429$, $x_8 = 0.79849$; $|A| = 42.821$. **44.** $x_1 = -1.2407$, $x_2 = -0.83809$, $x_3 = 0.84604$, $x_4 = 2.5237$, $x_5 = -0.25145$, $x_6 = -5.5907$, $x_7 = 2.3528$, $x_8 = -1.2883$, $x_9 = 6.4714$; $|A| = 445.14$. **45.** $x_1 = -0.029648$, $x_2 = 1.3788$, $x_3 = -0.054373$, $x_4 = 2.5869$, $x_5 = -1.3890$, $x_6 = 5.3499$, $x_7 = -2.3572$, $x_8 = -0.21915$, $x_9 = 2.7714$, $x_{10} = -2.5095$; $|A| = -290.58$.

46.

k	x_1	x_2	x_3
0	0.58849	0.83795	−0.20896
1	0.52148	0.75391	−0.22852
2	0.46863	0.68450	−0.25461
3	0.42626	0.62590	−0.28957
4	0.39201	0.57532	−0.33757
5	0.36450	0.53053	−0.40649
6	0.34322	0.48941	−0.51271
7	0.32908	0.44898	−0.69643
8	0.32740	0.40214	−1.0890
9	0.37500	0.30882	−2.5074
10	−0.13043	0.80435	8.2174
11	0.16045	0.46269	1.5560
12	0.18011	0.41088	0.85929
13	0.18128	0.38152	0.59360
14	0.17774	0.35963	0.45349
15	0.17265	0.34158	0.36696

47.

k	x_1	x_2	x_3	x_4
0.00	1.0000	0.0000	−1.0000	0.0000
0.01	1.0141	−0.021166	−0.97517	−0.025100
0.02	1.0254	−0.039437	−0.95326	−0.047393
0.03	1.0344	−0.055311	−0.93378	−0.067339
0.04	1.0414	−0.069182	−0.91633	−0.085304
0.05	1.0468	−0.081361	−0.90059	−0.10158
0.06	1.0509	−0.092099	−0.88631	−0.11640
0.07	1.0539	−0.10160	−0.87328	−0.12995
0.08	1.0558	−0.11004	−0.86134	−0.14240
0.09	1.0569	−0.11754	−0.85034	−0.15388
0.10	1.0573	−0.12424	−0.84016	−0.16449
0.11	1.0571	−0.13023	−0.83070	−0.17434
0.12	1.0563	−0.13560	−0.82189	−0.18351
0.13	1.0551	−0.14040	−0.81364	−0.19205
0.14	1.0534	−0.14472	−0.80590	−0.20004
0.15	1.0514	−0.14859	−0.79861	−0.20752

48.

k	x_1	x_2	x_3	x_4	x_5
0	−1.3621	−0.16734	0.97744	−0.20251	−1.1554
1	−1.6854	−0.41289	0.90938	−0.35205	−1.2969
2	−2.2314	−0.81039	0.75259	−0.65034	−1.5407
3	−3.4087	−1.6236	0.34827	−1.3816	−2.0802
4	−8.2628	−4.8124	−1.4771	−4.6644	−4.3594
5	12.226	8.2886	6.4367	9.6812	5.3812
6	2.9644	2.2414	2.8958	3.3416	1.0196
7	1.4547	1.1684	2.3228	2.3947	0.33679
8	0.81904	0.64609	2.0699	2.0558	0.071236
9	0.46203	0.29227	1.9061	1.9085	−0.060148
10	0.23190	0.010881	1.7714	1.8450	−0.13030
11	0.072784	−0.23193	1.6435	1.8241	−0.16689
12	−0.040413	−0.44979	1.5128	1.8262	−0.18310
13	−0.12058	−0.64819	1.3751	1.8403	−0.18641
14	−0.17510	−0.82901	1.2293	1.8601	−0.18162

49.

k	x_1	x_2	x_3	x_4	x_5	x_6
0	0.48110	0.44225	0.74246	0.78502	0.10840	1.0560
1	0.51052	0.37778	0.74921	0.74399	0.14914	1.0432
2	0.50208	0.35183	0.73886	0.69577	0.18960	1.0313
3	0.48706	0.33490	0.72639	0.64926	0.22660	1.0174
4	0.47052	0.32198	0.71416	0.60543	0.25993	1.0018

k	x_1	x_2	x_3	x_4	x_5	x_6
5	0.45396	0.31138	0.70283	0.56428	0.28977	0.98468
6	0.43791	0.30238	0.69260	0.52562	0.31640	0.96647
7	0.42258	0.29454	0.68350	0.48920	0.34008	0.94747
8	0.40803	0.28761	0.67549	0.45479	0.36108	0.92798
9	0.39429	0.28142	0.66851	0.42221	0.37966	0.90820
10	0.38133	0.27582	0.66250	0.39126	0.39605	0.88833
11	0.36911	0.27073	0.65737	0.36180	0.41045	0.86849
12	0.35759	0.26605	0.65307	0.33369	0.42307	0.84882
13	0.34673	0.26174	0.64950	0.30683	0.43408	0.82939
14	0.33648	0.25773	0.64661	0.28110	0.44364	0.81029

50. $x_1 = -0.14313$, $x_2 = 3.4371$, $x_3 = 0.35625$, $x_4 = 1.3242$, $x_5 = 1.1338$, $x_6 = 2.3030$, $x_7 = -1.3409$, $x_8 = 0.70942$. **51.** $x_1 = 20.737$, $x_2 = 22.256$, $x_3 = -14.311$, $x_4 = -1.5676$, $x_5 = -45.293$, $x_6 = 24.732$, $x_7 = 44.229$, $x_8 = -38.600$, $x_9 = -14.808$. **52.** $x_1 = 0.34816$, $x_2 = 0.40249$, $x_3 = -0.43917$, $x_4 = -1.0056$, $x_5 = -0.20730$, $x_6 = 0.21284$, $x_7 = -0.34594$, $x_8 = -0.59495$, $x_9 = -1.1252$, $x_{10} = -0.80465$.

53. a)

k	'inverse matrix'
0	$\begin{bmatrix} 1.0448 & 0.19190 & 0.085288 \\ 0.074627 & 1.0341 & -0.095949 \\ -0.29851 & 0.14925 & -1.0448 \end{bmatrix}$
1	$\begin{bmatrix} 0.94727 & 0.15625 & 0.087891 \\ 0.058594 & 0.93750 & -0.097656 \\ -0.30273 & 0.15625 & -1.1621 \end{bmatrix}$
2	$\begin{bmatrix} 0.86716 & 0.12915 & 0.092251 \\ 0.046125 & 0.85793 & -0.10148 \\ -0.31365 & 0.16605 & -1.3100 \end{bmatrix}$
3	$\begin{bmatrix} 0.80036 & 0.10791 & 0.098921 \\ 0.035971 & 0.79137 & -0.10791 \\ -0.33273 & 0.17986 & -1.5018 \end{bmatrix}$
4	$\begin{bmatrix} 0.74410 & 0.090744 & 0.10889 \\ 0.027223 & 0.73503 & -0.11797 \\ -0.36298 & 0.19964 & -1.7604 \end{bmatrix}$
5	$\begin{bmatrix} 0.69656 & 0.076336 & 0.12405 \\ 0.019084 & 0.68702 & -0.13359 \\ -0.41031 & 0.22901 & -2.1279 \end{bmatrix}$

k	0	1	2	3	4	5
M	3	4	5	6	7	10

53. b)

k	'inverse matrix'			
0.00	2.7292	−2.1421	2.2211	−2.0819
	−2.0819	2.7292	−2.1421	2.2211
	2.2211	−2.0819	2.7292	−2.1421
	−2.1421	2.2211	−2.0819	2.7292
0.01	2.5827	−1.9880	2.0674	−1.9272
	−1.9432	2.5742	−1.9923	2.0674
	2.0787	−1.9296	2.5742	−1.9880
	−2.0158	2.0787	−1.9432	2.5827
0.02	2.4533	−1.8526	1.9323	−1.7915
	−1.8211	2.4376	−1.8606	1.9323
	1.9532	−1.7958	2.4376	−1.8526
	−1.9047	1.9532	−1.8211	2.4533
0.03	2.3382	−1.7328	1.8125	−1.6715
	−1.7127	2.3162	−1.7439	1.8125
	1.8417	−1.6775	2.3162	−1.7328
	−1.8063	1.8417	−1.7127	2.3382
0.04	2.2350	−1.6261	1.7057	−1.5647
	−1.6160	2.2077	−1.6399	1.7057
	1.7421	−1.5720	2.2077	−1.6261
	−1.7184	1.7421	−1.6160	2.2350
0.05	2.1419	−1.5305	1.6099	−1.4690
	−1.5291	2.1101	−1.5465	1.6099
	1.6524	−1.4776	2.1101	−1.5305
	−1.6395	1.6524	−1.5291	2.1419
0.06	2.0575	−1.4443	1.5234	−1.3829
	−1.4506	2.0218	−1.4622	1.5234
	1.5713	−1.3925	2.0218	−1.4443
	−1.5682	1.5713	−1.4506	2.0575
0.07	1.9806	−1.3663	1.4449	−1.3050
	−1.3794	1.9415	−1.3859	1.4449
	1.4976	−1.3155	1.9415	−1.3663
	−1.5036	1.4976	−1.3794	1.9806
0.08	1.9102	−1.2954	1.3735	−1.2343
	−1.3144	1.8681	−1.3163	1.3735
	1.4303	−1.2455	1.8681	−1.2954
	−1.4447	1.4303	−1.3144	1.9102

k	'inverse matrix'			
0.09	1.8456	−1.2306	1.3082	−1.1698
	−1.2549	1.8009	−1.2528	1.3082
	1.3686	−1.1817	1.8009	−1.2306
	−1.3907	1.3686	−1.2549	1.8456
0.10	1.7859	−1.1713	1.2483	−1.1107
	−1.2003	1.7389	−1.1944	1.2483
	1.3118	−1.1231	1.7389	−1.1713
	−1.3412	1.3118	−1.2003	1.7859

k	0.00	0.01	0.02	0.03	0.04	0.05
M	15	14	13	13	12	12

k	0.06	0.07	0.08	0.09	0.10
M	12	11	11	11	10

53. c)

k	'inverse matrix'				
0	−0.40640	−0.63827	0.031779	−0.94433	−1.4829
	−0.67858	−0.39516	0.0010823	−0.071150	−0.86566
	−0.45387	−0.26430	−0.37294	−0.021178	−0.10655
	−0.41500	−0.24167	−0.34100	−0.76884	−0.56719
	−0.22041	−0.12835	−0.18111	−0.40833	−0.77970
1	−0.57209	−0.75772	−0.041110	−1.1690	−1.8438
	−0.77762	−0.48406	−0.024813	−0.21007	−1.1226
	−0.52424	−0.32633	−0.39335	−0.090632	−0.24063
	−0.53329	−0.33197	−0.40014	−0.92602	−0.79368
	−0.29746	−0.18517	−0.22320	−0.51652	−0.94522
2	−0.86323	−0.97362	−0.17166	−1.5646	−2.4745
	−0.95698	−0.63415	−0.086824	−0.45890	−1.5558
	−0.64318	−0.42621	−0.43382	−0.22431	−0.48074
	−0.73982	−0.49024	−0.49901	−1.2008	−1.2065
	−0.43450	−0.28792	−0.29307	−0.70526	−1.2377
3	−1.5133	−1.4652	−0.46618	−2.4460	−3.8739
	−1.3667	−0.96054	−0.25341	−1.0204	−2.4864
	−0.90036	−0.63278	−0.53497	−0.54336	−1.0194
	−1.1970	−0.84127	−0.71124	−1.8121	−2.1528
	−0.74228	−0.52168	−0.44105	−1.1237	−1.8937

k	'inverse matrix'				
4	−4.2624	−3.5684	−1.7157	−6.1590	−9.7617
	−3.1258	−2.3224	−1.0335	−3.4037	−6.3073
	−1.9648	−1.4598	−1.0008	−1.9502	−3.2881
	−3.1148	−2.3142	−1.5866	−4.3910	−6.2197
	−2.0458	−1.5200	−1.0421	−2.8840	−4.6769
5	7.4645	5.4441	3.6135	9.6426	15.300
	4.4273	3.4672	2.4184	6.7646	9.7761
	2.5391	1.9884	1.0666	4.1504	6.3436
	5.0330	3.9415	2.1142	6.6022	11.236
	3.5136	2.7516	1.4760	4.6091	7.2149
6	2.1989	1.4087	1.2178	2.5334	4.0310
	1.0514	0.86558	0.90171	2.1951	2.4879
	0.50746	0.41779	0.16558	1.4393	1.9963
	1.3634	1.1225	0.44485	1.6649	3.4271
	1.0157	0.83619	0.33140	1.2403	1.8820
7	1.3603	0.77291	0.83292	1.3913	2.2280
	0.52416	0.45259	0.67818	1.4624	1.2855
	0.17952	0.15501	0.040935	1.0241	1.2844
	0.77150	0.66616	0.17592	0.87870	2.2001
	0.61609	0.53197	0.14048	0.70169	1.0375

k	0	1	2	3	4	5	6	7
M	22	27	36	57	144	225	59	33

d)

k	'inverse matrix'					
0	8.2200	−5.3630	−3.5118	2.6735	1.3406	−1.3542
	−7.4114	5.2885	3.4631	−2.6363	−1.3220	1.3354
	3.7778	−2.6957	−1.3932	1.0606	0.53183	−0.53722
	2.2060	−1.5741	−0.81351	1.0518	0.52745	−0.53280
	−0.82031	0.58535	0.30251	−0.39114	0.085576	−0.086445
	−0.80782	0.57643	0.29791	−0.38518	0.084273	0.42249
1	3.6846	−2.3118	−1.4935	1.1522	0.57692	−0.55462
	−3.1948	2.4403	1.5765	−1.2162	−0.60897	0.58543
	1.6066	−1.2271	−0.42571	0.32843	0.16444	−0.15809
	0.83298	−0.63624	−0.22072	0.58506	0.29294	−0.28162
	−0.30929	0.23624	0.081954	−0.21724	0.17252	−0.16585
	−0.28986	0.22140	0.076807	−0.20359	0.16169	0.32765
2	2.3922	−1.4449	−0.92187	0.72035	0.35926	−0.32946
	−1.9968	1.6256	1.0371	−0.81040	−0.40417	0.37064
	0.99168	−0.80729	−0.15255	0.11921	0.059451	−0.054520
	0.45466	−0.37012	−0.069942	0.45333	0.22609	−0.20733
	−0.16815	0.13689	0.025868	−0.16766	0.19657	−0.18026
	−0.15033	0.12238	0.023126	−0.14989	0.17573	0.29967
3	1.7802	−1.0361	−0.65349	0.51690	0.25624	−0.22463
	−1.4319	1.2376	0.78056	−0.61741	−0.30606	0.26831
	0.70297	−0.60759	−0.024857	0.019661	0.0097465	−0.0085443
	0.28342	−0.24496	−0.010021	0.39190	0.19427	−0.17031
	−0.10419	0.090052	0.0036840	−0.14407	0.20708	−0.18154
	−0.089042	0.076961	0.0031485	−0.12312	0.17697	0.28538
4	1.4229	−0.79882	−0.49852	0.39893	0.19621	−0.16475
	−1.1039	1.0096	0.63007	−0.50420	−0.24799	0.20823
	0.53627	−0.49046	0.048507	−0.038816	−0.019092	0.016031
	0.18874	−0.17262	0.017072	0.35681	0.17550	−0.14736
	−0.068842	0.062961	−0.0062269	−0.13015	0.21231	−0.17827
	−0.056352	0.051538	−0.0050971	−0.10653	0.17379	0.27602
5	1.1885	0.64422	−0.39815	0.32214	0.15697	−0.12647
	−0.89027	0.85895	0.53086	−0.42953	−0.20930	0.16862
	0.42829	−0.41323	0.095766	−0.077485	−0.037756	0.030419
	0.13033	−0.12575	0.029143	0.33443	0.16296	−0.13129
	−0.047096	0.045439	−0.010531	−0.12085	0.21486	−0.17311
	−0.036990	0.035689	−0.0082710	−0.094916	0.16876	0.26889

k	0	1	2	3	4	5
M	98	46	31	24	20	18

53. e)

$$A^{-1} = \begin{bmatrix} 1.4456 & -0.54225 & -1.4585 & -0.21416 & -1.0789 & 0.70487 & 0.87256 & 1.7477 \\ -2.1940 & 1.2995 & 3.8486 & 1.2983 & 1.6163 & -1.4819 & -2.2078 & -3.6443 \\ 1.1128 & -0.32773 & -0.81368 & -0.35089 & -0.35776 & 0.52875 & 0.73335 & 0.54822 \\ -0.97552 & 0.80311 & 0.30382 & 0.42994 & 0.29879 & -0.22775 & -0.52671 & -0.64602 \\ -1.1485 & 0.33914 & 0.22377 & 0.69301 & 0.89808 & -0.034526 & -0.55240 & -0.64152 \\ -2.1359 & 1.2172 & 1.7285 & 0.21971 & 1.3465 & -1.1826 & -1.4210 & -1.9441 \\ 2.7184 & -1.6825 & -2.8048 & 0.55434 & -2.0947 & 1.2012 & 2.3483 & 3.6613 \\ -0.28900 & 0.34822 & -0.53703 & 0.38801 & -0.16070 & 0.18549 & 0.10611 & 0.44319 \end{bmatrix};$$

$M = 78.$

53. f)

$$A^{-1} = \begin{bmatrix} 4.3406 & -3.5559 & 1.7190 & 5.1942 & -3.5029 & 0.91309 & -7.8436 & 4.7957 & 2.7324 \\ 4.8397 & -3.6090 & 1.2466 & 5.5134 & -4.3003 & 0.74500 & -9.6549 & 5.4376 & 2.2941 \\ -3.0018 & 2.7466 & -0.53748 & -3.3574 & 3.0451 & -0.98114 & 6.8794 & -3.4812 & -1.4426 \\ -0.12062 & -0.0030265 & 0.22618 & -0.33214 & 1.1333 & -0.42615 & 0.93232 & -0.71810 & -0.16297 \\ -10.137 & 7.8772 & -3.8317 & -11.292 & 8.0717 & -2.3555 & 19.565 & -11.690 & -5.5286 \\ 6.0338 & -3.6745 & 2.2069 & 5.6407 & -3.5084 & 1.2878 & -10.080 & 5.4229 & 2.8261 \\ 9.7492 & -6.5359 & 3.8063 & 10.535 & -6.9177 & 1.7105 & -17.769 & 10.696 & 5.8418 \\ -9.9030 & 6.4469 & -3.5860 & -9.2722 & 5.8302 & -1.6566 & 16.527 & -9.6053 & -5.1826 \\ -3.9447 & 2.6704 & -1.8063 & -3.6227 & 2.8649 & -0.58946 & 6.6354 & -3.5763 & -2.0717 \end{bmatrix};$$

$M = 194.$

53. g) $A^{-1} =$

$$\begin{bmatrix} 0.94609 & 3.5321 & -2.0707 & -0.68384 & -0.26044 & 1.0390 & 0.79193 & 0.16170 & -2.2671 & -1.6107 \\ -0.37684 & -1.3169 & 0.66837 & 0.35973 & -0.026334 & -0.35300 & -0.51867 & -0.034309 & 0.78098 & 0.32909 \\ 0.31747 & 1.6138 & -0.54389 & -0.31931 & 0.047734 & 0.29099 & 0.47001 & 0.032592 & -0.63987 & -0.73714 \\ -0.79753 & -4.2037 & 2.5122 & 1.1695 & 0.31958 & -1.1300 & -1.0723 & -0.086568 & 2.6470 & 1.8607 \\ 0.16145 & 0.92588 & -0.56460 & 0.0057578 & 0.17985 & 0.30921 & 0.083989 & -0.0095503 & -0.45960 & -0.38111 \\ 0.13191 & 0.085685 & -0.16129 & -0.0064821 & 0.097065 & 0.28474 & -0.069109 & -0.13731 & -0.21155 & -0.27165 \\ 0.41427 & 2.2129 & -1.0078 & -0.42029 & -0.15788 & 0.73131 & 0.75338 & 0.15453 & -1.2928 & -0.96824 \\ -0.10564 & -0.82527 & 0.46243 & 0.40277 & 0.095733 & -0.25425 & -0.061401 & 0.19512 & 0.54929 & 0.26236 \\ -0.18132 & -0.58997 & 0.54063 & 0.13278 & 0.34157 & -0.17638 & -0.11651 & 0.20172 & 0.80995 & 0.40137 \\ -0.81667 & -4.3942 & 2.4067 & 1.0026 & 0.30884 & -1.0173 & -1.0799 & -0.22987 & 2.8845 & 1.9106 \end{bmatrix};$$

$M = 116.$

54.

k	x_1	x_2	x_3
0.0	1.0000	1.0000	1.0000
0.1	0.96980	0.97924	0.98418
0.2	0.94205	0.95910	0.96839
0.3	0.91637	0.93960	0.95273
0.4	0.89248	0.92077	0.93728
0.5	0.87013	0.90260	0.92208
0.6	0.84914	0.88506	0.90717
0.7	0.82937	0.86815	0.89256
0.8	0.81067	0.85183	0.87829
0.9	0.79295	0.83609	0.86435
1.0	0.77612	0.82090	0.85075
1.1	0.76009	0.80623	0.83748
1.2	0.74481	0.79206	0.82455
1.3	0.73020	0.77838	0.81196
1.4	0.71622	0.76515	0.79969
1.5	0.70283	0.75236	0.78774

55.

k	x_1	x_2	x_3	x_4
0.0	1.0000	1.0000	1.0000	1.0000
0.1	0.55226	1.2688	1.1142	0.93150
0.2	0.41943	1.3475	1.1480	0.91122
0.3	0.35548	1.3846	1.1642	0.90149
0.4	0.31773	1.4060	1.1737	0.89579
0.5	0.29270	1.4196	1.1799	0.89204
0.6	0.27483	1.4290	1.1844	0.88939
0.7	0.26137	1.4357	1.1876	0.88742
0.8	0.25082	1.4406	1.1902	0.88590
0.9	0.24229	1.4444	1.1922	0.88470
1.0	0.23523	1.4472	1.1938	0.88373
1.1	0.22926	1.4494	1.1951	0.88293
1.2	0.22412	1.4511	1.1962	0.88226
1.3	0.21964	1.4524	1.1972	0.88170
1.4	0.21568	1.4535	1.1979	0.88123
1.5	0.21214	1.4542	1.1986	0.88083

56.

k	x_1	x_2	x_3	x_4	x_5
0.0	1.0102	−1.0256	0.010494	−0.97569	0.99776
0.1	1.0117	−1.0293	0.011994	−0.97259	0.99768
0.2	1.0132	−1.0330	0.013507	−0.96950	0.99760
0.3	1.0147	−1.0368	0.015033	−0.96643	0.99755
0.4	1.0162	−1.0406	0.016572	−0.96337	0.99751
0.5	1.0176	−1.0444	0.018124	−0.96032	0.99748
0.6	1.0191	−1.0482	0.019690	−0.95729	0.99748
0.7	1.0206	−1.0520	0.021270	−0.95428	0.99748
0.8	1.0221	−1.0559	0.022863	−0.95128	0.99750
0.9	1.0236	−1.0598	0.024471	−0.94829	0.99754
1.0	1.0251	−1.0638	0.026093	−0.94532	0.99759
1.1	1.0266	−1.0677	0.027730	−0.94237	0.99766
1.2	1.0281	−1.0717	0.029381	−0.93942	0.99775
1.3	1.0296	−1.0757	0.031047	−0.93649	0.99785
1.4	1.0311	−1.0797	0.032729	−0.93358	0.99796
1.5	1.0326	−1.0838	0.034425	−0.93068	0.99809

57.

k	x_1	x_2	x_3	x_4	x_5	x_6
0	0.44481	0.19411	1.1946	0.48086	1.7025	− 2.4568
1	0.16887	1.0983	1.3060	0.22309	0.64059	− 2.1205
2	0.016994	2.8308	1.2979	− 0.11997	0.13717	− 3.0637
3	−7.8966	186.57	11.033	−30.571	−27.962	−141.89
4	0.21302	− 3.5471	0.77004	0.85880	0.94945	2.1148
5	0.11168	− 1.8295	0.79108	0.54296	0.65583	0.91919
6	0.060161	− 1.2441	0.76432	0.41586	0.54829	0.55702
7	0.019658	− 0.94226	0.73341	0.33532	0.49336	0.39418
8	−0.019490	− 0.75515	0.70626	0.27151	0.46387	0.30813
9	−0.062966	− 0.62607	0.68570	0.21242	0.45152	0.25982
10	−0.11724	− 0.53013	0.67418	0.14957	0.45494	0.23401
11	−0.19359	− 0.45408	0.67617	0.072123	0.47839	0.22534
12	−0.31827	− 0.38885	0.70277	− 0.042172	0.53682	0.23589
13	−0.57595	− 0.32335	0.79106	− 0.26266	0.68304	0.28383
14	−1.4949	− 0.21210	1.1746	− 1.0186	1.2525	0.49669
15	5.2270	− 0.72506	− 1.8237	4.4293	− 3.0387	− 1.1546

58. $x_1 = 1.1184$, $x_2 = 0.22640$, $x_3 = 0.30246$, $x_4 = 0.68737$, $x_5 = 0.22233$, $x_6 = -0.61787$, $x_7 = 1.1607$. **59.** $x_1 = 1.0469$, $x_2 = 1.0555$, $x_3 = 0.91398$, $x_4 = 0.94450$, $x_5 = 1.0726$, $x_6 = 0.93365$, $x_7 = 0.87229$, $x_8 = 0.95200$. **60.** $x_1 = 0.83650$, $x_2 = 0.96636$, $x_3 = 1.2858$, $x_4 = 1.0348$, $x_5 = 0.95991$, $x_6 = 1.1487$, $x_7 = 0.77761$, $x_8 = 0.99853$, $x_9 = 0.94682$. **61.** $x_1 = -13.710$, $x_2 = -14.117$, $x_3 = 9.9198$, $x_4 = 2.0412$, $x_5 = 18.977$. $x_6 = -15.580$, $x_7 = 4.3841$, $x_8 = -6.5642$, $x_9 = -13.350$, $x_{10} = 24.880$.

62.

k	x_1	x_2	x_3
0.0	4.0356	3.2017	5.6071
0.5	3.7617	3.1635	5.4450
1.0	3.5173	3.1308	5.2850
1.5	3.2988	3.1019	5.1283
2.0	3.1031	3.0754	4.9760
2.5	2.9274	3.0505	4.8286
3.0	2.7691	3.0264	4.6865
3.5	2.6262	3.0028	4.5498
4.0	2.4968	2.9793	4.4186
4.5	2.3793	2.9557	4.2929
5.0	2.2723	2.9319	4.1727
5.5	2.1746	2.9079	4.0576
6.0	2.0851	2.8835	3.9476
6.5	2.0028	2.8589	3.8425
7.0	1.9271	2.8339	3.7420
7.5	1.8572	2.8088	3.6460

63.

k	x_1	x_2	x_3	x_4
0	1.2924	0.68463	1.6282	0.69091
1	1.0052	0.72150	1.2684	0.74862
2	0.85736	0.67396	1.0717	0.71664
3	0.75696	0.61859	0.93593	0.67131
.4	0.68119	0.56739	0.83360	0.62634
5	0.62084	0.52227	0.75269	0.58493
6	0.57114	0.48297	0.68671	0.54765
7	0.52928	0.44872	0.63170	0.51430
8	0.49341	0.41874	0.58504	0.48446
9	0.46228	0.39236	0.54492	0.45771
10	0.43496	0.36900	0.51003	0.43363
11	0.41077	0.34819	0.47938	0.41188
12	0.38918	0.32956	0.45225	0.39214
13	0.36979	0.31279	0.42804	0.37418
14	0.35227	0.29761	0.40632	0.35775
15	0.33636	0.28382	0.38670	0.34269

64.

k	x_1	x_2	x_3	x_4	x_5
0	0.0030579	0.068616	−0.071812	0.21482	−0.019594
1	0.0044020	0.058483	−0.058615	0.17963	−0.016384
2	0.0049936	0.050974	−0.049420	0.15436	−0.014078
3	0.0052152	0.045182	−0.042667	0.13533	−0.012341
4	0.0052459	0.040577	−0.037507	0.12048	−0.010985
5	0.0051760	0.036827	−0.033442	0.10858	−0.0098980
6	0.0050532	0.033713	−0.030160	0.098811	−0.0090066
7	0.0049040	0.031086	−0.027457	0.090660	−0.0082626
8	0.0047435	0.028840	−0.025193	0.083751	−0.0076321
9	0.0045801	0.026896	−0.023270	0.077822	−0.0070909
10	0.0044188	0.025199	−0.021617	0.072677	−0.0066215
11	0.0042623	0.023703	−0.020182	0.068170	−0.0062103
12	0.0041121	0.022376	−0.018923	0.064190	−0.0058472
13	0.0039689	0.021189	−0.017811	0.060650	−0.0055242
14	0.0038329	0.020122	−0.016822	0.057479	−0.0052350
15	0.0037042	0.019157	−0.015936	0.054624	−0.0049746

65.

k	x_1	x_2	x_3	x_4	x_5	x_6
0	−0.13242	0.15271	−0.033703	0.12132	0.029552	0.13399
1	−0.097041	0.12466	−0.023556	0.098831	0.031147	0.10829
2	−0.075697	0.10579	−0.017578	0.083435	0.030594	0.090994
3	−0.061600	0.092096	−0.013737	0.072218	0.029295	0.078532
4	−0.051681	0.081649	−0.011110	0.063674	0.027775	0.069109
5	−0.044367	0.073389	−0.0092259	0.056946	0.026243	0.061726
6	−0.038776	0.066680	−0.0078228	0.051509	0.024781	0.055781
7	−0.034377	0.061116	−0.0067464	0.047023	0.023422	0.050889
8	−0.030835	0.056423	−0.0059001	0.043257	0.022171	0.046792
9	−0.027928	0.052407	−0.0052208	0.040052	0.021026	0.043308
10	−0.025502	0.048931	−0.0046661	0.037289	0.019979	0.040310
11	−0.023450	0.045892	−0.0042062	0.034884	0.019022	0.037702
12	−0.021693	0.043212	−0.0038200	0.032771	0.018145	0.035413
13	−0.020173	0.040829	−0.0034920	0.030899	0.017340	0.033386
14	−0.018846	0.038697	−0.0032105	0.029230	0.016600	0.031580
15	−0.017678	0.036778	−0.0029668	0.027733	0.015918	0.029960

66. $x_1 = 0.43093$, $x_2 = 0.41283$, $x_3 = 0.40341$, $x_4 = 0.42251$, $x_5 = 0.45070$, $x_6 = 0.41265$, $x_7 = 0.43093$. **67.** $x_1 = 0.18306$, $x_2 = 0.062374$, $x_3 = 0.076600$, $x_4 = 0.32525$, $x_5 = 0.41374$, $x_6 = 0.20347$, $x_7 = 0.19385$, $x_8 = 0.24750$. **68.** $x_1 = 0.31509$, $x_2 = 0.087364$, $x_3 = 0.084250$, $x_4 = 0.22534$, $x_5 = 0.21509$, $x_6 = 0.27540$, $x_7 = 0.026167$, $x_8 = 0.28279$, $x_9 = 0.38181$. **69.** $x_1 = 0.12754$, $x_2 = 0.14324$, $x_3 = 0.078535$, $x_4 = 0.18392$, $x_5 = 0.075688$, $x_6 = 0.13889$, $x_7 = -0.066209$, $x_8 = 0.37060$, $x_9 = 0.15760$, $x_{10} = 0.29872$.

70. $\lambda_1 = -0.951374$, $\lambda_2 = -0.750564$, $\lambda_3 = 0.669057$, $\lambda_4 = 0.826996$;

$$X_1 = \begin{bmatrix} -0.784746 \\ 0.269371 \\ 0.437805 \\ 1.00000 \end{bmatrix}, \quad X_2 = \begin{bmatrix} 0.437805 \\ 1.00000 \\ 0.784745 \\ -0.269371 \end{bmatrix}, \quad X_3 = \begin{bmatrix} 1.00000 \\ 0.220158 \\ -0.512420 \\ 0.949781 \end{bmatrix},$$

$$X_4 = \begin{bmatrix} 0.512420 \\ -0.949781 \\ 1.00000 \\ 0.220158 \end{bmatrix}.$$

71. $\lambda_1 = -0.941150$, $\lambda_2 = -0.747184$, $\lambda_3 = 0.674904$, $\lambda_4 = 0.835388$;

$$X_1 = \begin{bmatrix} 1.00000 \\ 0.215129 \\ -0.445661 \\ -0.864259 \end{bmatrix}, \quad X_2 = \begin{bmatrix} 0.445662 \\ 0.864259 \\ 1.00000 \\ 0.215130 \end{bmatrix}, \quad X_3 = \begin{bmatrix} 0.684377 \\ -0.086491 \\ -0.445381 \\ 1.00000 \end{bmatrix},$$

$$X_4 = \begin{bmatrix} -0.445380 \\ 1.00000 \\ -0.684377 \\ 0.086490 \end{bmatrix}.$$

72. $\lambda_1 = -0.943318$, $\lambda_2 = -0.765617$, $\lambda_3 = 0.675453$, $\lambda_4 = 0.844798$;

$$X_1 = \begin{bmatrix} 1.00000 \\ -0.483706 \\ 0.210941 \\ -0.580366 \end{bmatrix}, \quad X_2 = \begin{bmatrix} -0.210941 \\ 0.580366 \\ 1.00000 \\ -0.483706 \end{bmatrix}, \quad X_3 = \begin{bmatrix} 0.662583 \\ 0.352619 \\ 0.418824 \\ 1.00000 \end{bmatrix},$$

$$X_4 = \begin{bmatrix} 0.418823 \\ 1.00000 \\ -0.662583 \\ -0.352619 \end{bmatrix}.$$

73. $\lambda_1 = -1.00252$, $\lambda_2 = -0.803278$, $\lambda_3 = 0.479339$, $\lambda_4 = 0.840311$;

$$X_1 = \begin{bmatrix} 1.00000 \\ 0.489359 \\ -0.600155 \\ -0.864886 \end{bmatrix}, \quad X_2 = \begin{bmatrix} 0.717823 \\ 0.773454 \\ 1.00000 \\ 0.573676 \end{bmatrix}, \quad X_3 = \begin{bmatrix} 0.479139 \\ -0.172999 \\ -0.783806 \\ -1.00000 \end{bmatrix},$$

$$X_4 = \begin{bmatrix} -0.614506 \\ 1.00000 \\ -0.414239 \\ 0.142749 \end{bmatrix}.$$

74.

k	λ_i	x_1	x_2	x_3
0.0	0.019910	1.00000	0.213838	−0.612831
	0.163329	−0.511996	1.00000	−0.486525
	1.48476	0.458585	0.721319	1.00000
0.1	0.031944	1.00000	0.242471	−0.677864
	0.295417	−0.528986	1.00000	−0.422673
	1.64064	0.509968	0.692439	1.00000
0.2	0.041702	1.00000	0.240105	−0.713664
	0.424510	−0.510377	1.00000	−0.378710
	1.80179	0.554752	0.661843	1.00000
0.3	0.049917	1.00000	0.230694	−0.739289
	0.550549	−0.484404	1.00000	−0.343181
	1.96753	0.593766	0.630804	1.00000
0.4	0.056947	1.00000	0.219563	−0.759572
	0.673771	−0.457369	1.00000	−0.313080
	2.13728	0.627788	0.600211	1.00000
0.5	0.063030	1.00000	0.208371	−0.776429
	0.794462	−0.431270	1.00000	−0.287083
	2.31051	0.657522	0.570652	1.00000
0.6	0.068344	1.00000	0.197701	−0.790834
	0.912907	−0.406799	1.00000	−0.264401
	2.48675	0.683585	0.542483	1.00000

75.

k	λ_i	x_1	x_2	x_3
0.0	1.1851	1.0000	−0.58777	−0.22713
	1.7599	−0.31603	−0.92411	1.0000
	4.5550	0.71871	0.83634	1.0000
0.1	1.2679	1.0000	−0.50555	−0.29687
	1.7366	0.21462	1.0000	−0.97999
	4.7955	0.71475	0.82659	1.0000
0.2	1.3391	1.0000	−0.38525	−0.39641
	1.7249	0.04746	1.0000	−0.85212
	5.0360	0.71168	0.81835	1.0000
0.3	1.3899	1.0000	−0.21934	−0.53133
	1.7334	−0.15377	1.0000	−0.70221
	5.2767	0.70927	0.81127	1.0000
0.4	1.4131	1.0000	−0.027009	−0.68560
	1.7694	−0.35353	1.0000	−0.55505
	5.5174	0.70735	0.80512	1.0000
0.5	1.4112	1.0000	0.15083	−0.82642
	1.8305	−0.51268	1.0000	−0.43785
	5.7583	0.70581	0.79971	1.0000

Continuation of 75

k	λ_i	x_1	x_2	x_3
0.6	1.3931	1.0000	0.29264	−0.93718
	1.9077	−0.62495	1.0000	−0.35459
	5.9992	0.70456	0.79491	1.0000
0.7	1.3662	−0.98036	−0.39244	1.0000
	1.9937	−0.70257	1.0000	−0.29633
	6.2401	0.70355	0.79062	1.0000
0.8	1.3344	−0.92416	−0.44560	1.0000
	2.0845	−0.75749	1.0000	−0.25445
	6.4811	0.70272	0.78676	1.0000
0.9	1.2998	−0.88547	−0.48306	1.0000
	2.1781	−0.79767	1.0000	−0.22326
	6.7221	0.70204	0.78326	1.0000
1.0	1.2634	−0.85752	−0.51080	1.0000
	2.2735	−0.82801	1.0000	−0.19923
	6.9631	0.70149	0.78007	1.0000
1.1	1.2258	−0.83652	−0.53216	1.0000
	2.3700	−0.85156	1.0000	−0.18019
	7.2042	0.70103	0.77716	1.0000
1.2	1.1874	−0.82026	−0.54912	1.0000
	2.4673	−0.87027	1.0000	−0.16472
	7.4453	0.70065	0.77448	1.0000
1.3	1.1484	−0.80734	−0.56293	1.0000
	2.5652	−0.88542	1.0000	−0.15191
	7.6864	−0.70035	0.77201	1.0000
1.4	1.1090	−0.79686	−0.57440	1.0000
	2.6634	−0.89790	1.0000	−0.14110
	7.9276	0.70010	0.76972	1.0000
1.5	1.0693	−0.78821	−0.58408	1.0000
	2.7620	−0.90833	1.0000	−0.13187
	8.1687	0.69989	0.76760	1.0000

76. k	λ_i	x_1	x_2	x_3	x_4
0	−2.673	1.000	0.1685	−0.3785	−0.8512
	−0.9667	0.1036	−0.7177	1.000	−0.4649
	0.1358	−0.5310	1.000	0.4763	−0.6377
	7.504	1.000	0.7238	0.8525	0.9391
1	−3.008	1.000	0.1460	−0.4932	−0.7779
	0.08728	0.1011	−0.6078	1.000	−0.6181
	1.234	−0.3900	1.000	0.3257	−0.5202
	9.686	1.000	0.5889	0.8046	0.8860
2	−3.380	1.000	0.1232	−0.5447	−0.7479
	1.123	0.08639	−0.5368	1.000	−0.7013
	2.328	−0.3154	1.000	0.2541	−0.4420
	11.93	1.000	0.4949	0.7772	0.8526
3	−3.767	1.000	0.1059	−0.5750	−0.7328
	2.147	0.07353	−0.4872	1.000	−0.7547
	3.408	−0.2669	1.000	0.2142	−0.3877
	14.21	1.000	0.4257	0.7601	0.8297
4	−4.162	1.000	0.09271	−0.5953	−0.7243
	3.164	0.06343	−0.4509	1.000	−0.7920
	4.474	−0.2320	1.000	0.1898	−0.3483
	16.52	1.000	0.3730	0.7487	0.8131
5	−4.563	1.000	0.08239	−0.6099	−0.7190
	4.176	0.05551	−0.4233	1.000	−0.8196
	5.530	−0.2054	1.000	0.1738	−0.3184
	18.86	1.000	0.3315	0.7408	0.8004
6	−4.966	1.000	0.07413	−0.6211	−0.7156
	5.186	0.04926	−0.4017	1.000	−0.8407
	6.576	−0.1844	1.000	0.1626	−0.2952
	21.20	1.000	0.2981	0.7350	0.7904
7	−5.372	1.000	0.06737	−0.6299	−0.7132
	6.193	0.04418	−0.3843	1.000	−0.8575
	7.616	−0.1673	1.000	0.1546	−0.2766
	23.56	1.000	0.2707	0.7307	0.7824
8	−5.779	1.000	0.06174	−0.6370	−0.7116
	7.199	0.04004	−0.3701	1.000	−0.8710
	8.649	−0.1531	1.000	0.1486	−0.2614
	25.93	1.000	0.2478	0.7274	0.7757

Continuation of 76

k	λ_i	x_1	x_2	x_3	x_4
9	−6.187	1.000	0.05698	−0.6429	−0.7103
	8.204	0.03657	−0.3583	1.000	−0.8823
	9.678	−0.1412	1.000	0.1439	−0.2488
	28.30	1.000	0.2285	0.7248	0.7702
10	−6.596	1.000	0.05290	−0.6479	−0.7094
	9.209	0.03366	−0.3482	1.000	−0.8917
	10.70	−0.1309	1.000	0.1403	−0.2381
	30.68	1.000	0.2118	0.7226	0.7654
11	−7.006	1.000	0.04936	−0.6521	−0.7087
	10.21	0.03119	−0.3397	1.000	0.8997
	11.73	−0.1221	1.000	0.1374	−0.2290
	33.07	1.000	0.1975	0.7209	0.7614
12	−7.417	1.000	0.04627	−0.6558	−0.7082
	11.22	0.02901	−0.3322	1.000	−0.9067
	12.74	−0.1144	1.000	0.1350	−0.2211
	35.46	1.000	0.1849	0.7195	0.7578
13	−7.827	1.000	0.04354	−0.6590	−0.7078
	12.22	0.02711	−0.3255	1.000	−0.9128
	13.76	−0.1076	1.000	0.1330	−0.2143
	37.85	1.000	0.1738	0.7183	0.7547
14	−8.239	1.000	0.04112	−0.6618	−0.7075
	13.22	0.02550	−0.3199	1.000	−0.9180
	14.78	−0.1015	1.000	0.1313	−0.2082
	40.24	1.000	0.1639	0.7173	0.7520
15	−8.650	1.000	0.03895	−0.6643	−0.7072
	14.22	0.02398	−0.3147	1.000	−0.9228
	15.79	−0.09613	1.000	0.1298	−0.2028
	42.64	1.000	0.1551	0.7165	0.7496

77.

k	λ_i	x_1	x_2	x_3	x_4
0	−13.159	−0.84138	0.41610	1.0000	−0.77176
	0.18295	0.01823	1.0000	−0.72458	−0.41959
	12.732	1.0000	−0.24759	0.22401	−0.93345
	30.244	0.82045	1.0000	0.91783	0.83397
1	−13.633	−0.78161	0.37810	1.0000	−0.72358
	0.68407	0.057335	1.0000	−0.67755	−0.47578
	13.050	1.0000	−0.29509	0.26181	−0.87257
	32.899	0.83349	1.0000	0.91944	0.89290

Continuation of 77

k	λ_i	x_1	x_2	x_3	x_4
2	−14.163	−0.73468	0.35001	1.0000	−0.68691
	1.1144	0.10455	1.0000	−0.63760	−0.53049
	13.450	1.0000	−0.34994	0.29547	−0.81771
	35.599	0.84926	1.0000	0.92230	0.94389
3	−14.736	−0.69685	0.32888	1.0000	−0.65845
	1.4629	0.15740	1.0000	−0.60366	−0.58388
	13.941	1.0000	−0.40900	0.32582	−0.76778
	38.332	0.86650	1.0000	0.92604	0.98885
4	−15.343	−0.66564	0.31277	1.0000	−0.63599
	1.7257	0.21357	1.0000	−0.57484	−0.63559
	14.524	1.0000	−0.46966	0.35315	−0.72231
	41.093	0.85942	0.97176	0.90411	1.0000

78.

k	λ_i	x_1	x_2	x_3	x_4	x_5
0	−14.863	−0.34940	−0.54920	1.0000	−0.69605	0.37519
	− 8.6685	1.0000	−0.54430	0.057104	−0.36491	−0.69466
	− 2.4567	0.30142	−0.48710	−0.87732	−0.48838	1.0000
	3.6874	−0.090926	−0.97566	0.085198	1.0000	0.11528
	17.301	1.0000	0.55592	0.68707	0.48170	0.80740
1	−14.775	−0.36845	−0.53690	1.0000	−0.68642	0.38383
	− 8.6124	1.0000	−0.55660	0.076822	−0.37894	−0.69648
	− 2.3329	0.30342	−0.49116	−0.87418	−0.49306	1.0000
	3.6870	−0.090926	−0.97640	0.084550	1.0000	0.11499
	17.533	1.0000	0.55435	0.68688	0.48135	0.80664
2	−14.691	−0.38757	−0.52462	1.0000	−0.67682	0.39265
	− 8.5525	1.0000	−0.56914	0.097128	−0.39322	−0.69853
	− 2.2088	0.30549	−0.49528	−0.87103	−0.49783	1.0000
	3.6866	−0.090911	−0.97714	0.083905	1.0000	0.11473
	17.765	1.0000	0.55280	0.68669	0.48099	0.80589
3	−14.611	−0.40674	−0.51238	1.0000	−0.66725	0.40164
	− 8.4889	1.0000	−0.58189	0.11800	−0.40773	−0.70082
	− 2.0842	0.30762	−0.49949	−0.86788	−0.50271	1.0000
	3.6863	−0.090882	−0.97788	0.083261	1.0000	0.11449
	17.998	1.0000	0.55127	0.68651	0.48062	0.80515
4	−14.536	−0.42590	−0.50019	1.0000	−0.65775	0.41077
	− 8.4213	1.0000	−0.59484	0.13943	−0.42244	−0.70336
	− 1.9591	0.30982	−0.50376	−0.86472	−0.50767	1.0000
	3.6859	−0.090839	−0.97863	0.082618	1.0000	0.11427
	18.23	1.0000	0.54977	0.68634	0.48024	0.80442

Continuation of 78

k	λ_i	x_1	x_2	x_3	x_4	x_5
5	−14.464	−0.44504	−0.48809	1.0000	−0.64831	0.42002
	− 8.3500	1.0000	−0.60796	0.16139	−0.43732	−0.70617
	− 1.8336	0.31208	−0.50811	−0.86155	−0.51274	1.0000
	3.6856	−0.090784	−0.97938	0.081974	1.0000	0.11407
	18.4625	1.0000	0.54829	0.68617	0.47984	0.80371
6	−14.398	−0.46411	−0.47607	1.0000	−0.63895	0.42937
	− 8.2748	1.0000	−0.62124	0.18386	−0.45235	−0.70926
	− 1.7076	0.31441	−0.51253	−0.85837	−0.51791	1.0000
	3.6852	−0.090715	−0.98013	0.081329	1.0000	0.11390
	18.695	1.0000	0.54683	0.68600	0.47944	0.80300
7	−14.335	−0.48309	−0.46418	1.0000	−0.62969	0.43879
	− 8.1959	1.0000	−0.63464	0.20682	−0.46751	−0.71265
	− 1.5812	0.31679	−0.51702	−0.85518	−0.52318	1.0000
	3.6848	−0.090633	−0.98090	0.080681	1.0000	0.11375
	18.927	1.0000	0.54540	0.68583	0.47902	0.80231
8	−14.277	−0.50192	−0.45241	1.0000	−0.62053	0.44826
	− 8.1134	1.0000	−0.64815	0.23022	−0.48277	−0.71633
	− 1.4543	0.31924	−0.52157	−0.85197	−0.52856	1.0000
	3.6845	−0.090537	−0.98168	0.080029	1.0000	0.11363
	19.160	1.0000	0.54398	0.68568	0.47859	0.80162
9	−14.222	−0.52060	−0.44079	1.0000	−0.61149	0.45775
	− 8.0272	1.0000	−0.66174	0.25406	−0.49809	−0.72033
	− 1.3269	0.32175	−0.52620	−0.84875	−0.53404	1.0000
	3.6841	−0.090429	−0.98247	0.079373	1.0000	0.11353
	19.392	1.0000	0.54258	0.68552	0.47816	0.80095
10	−14.172	−0.53907	−0.42934	1.0000	−0.60257	0.46724
	− 7.9375	1.0000	−0.67538	0.27830	−0.51346	−0.72466
	− 1.1990	0.32433	−0.53089	−0.84552	−0.53964	1.0000
	3.6838	−0.090308	−0.98328	0.078711	1.0000	0.11346
	19.625	1.0000	0.54121	0.68537	0.47771	0.80028

79.

k	λ_i	x_1	x_2	x_3	x_4	x_5	x_6
0	−2.5017	−0.6436	0.6959	−0.6497	0.8199	−0.5526	1.0000
	−0.4250	−0.5617	−0.3242	−0.5063	−0.0285	1.0000	0.1111
	0.7553	0.0926	1.0000	−0.0977	0.3885	0.4248	−0.7836
	2.5103	0.0724	−0.2864	0.9150	1.0000	0.4120	0.2481
	3.1846	1.0000	−0.1965	−0.6267	0.3263	0.1680	0.1984
	5.1765	0.4234	0.8237	0.5048	−0.7634	0.6276	1.0000

Continuation of 79

k	λ_i	x_1	x_2	x_3	x_4	x_5	x_6
1	−2.2162	−0.7084	0.6031	−0.6225	0.7480	−0.4676	1.0000
	0.0324	−0.5373	−0.4386	−0.4491	−0.1045	1.0000	0.1501
	1.2171	0.1094	1.0000	−0.2247	0.4468	0.5543	−0.7405
	3.0129	−0.0174	−0.2699	1.0000	0.9867	0.3917	0.2181
	3.8070	1.0000	−0.3676	−0.6456	0.4822	0.1041	0.2161
	5.8468	0.5783	0.7351	0.4119	−0.6649	0.5986	1.0000
2	−1.9686	−0.7590	0.5162	−0.5892	0.6777	−0.3950	1.0000
	0.4832	−0.5144	−0.5398	−0.3901	−0.1775	1.0000	0.1737
	1.6803	0.1155	1.0000	−0.3429	0.4996	0.6761	−0.7021
	3.5090	−0.0637	−0.2299	1.0000	0.9202	0.3652	0.1801
	4.4145	1.0000	−0.5382	−0.6907	0.6318	0.0296	0.2134
	6.5816	0.6997	0.6394	0.3291	−0.5671	0.5591	1.0000
3	−1.7563	−0.7983	0.4393	−0.5532	0.6125	−0.3351	1.0000
	0.9329	−0.4912	−0.6274	−0.3320	−0.2454	1.0000	0.1852
	2.1480	0.1151	1.0000	−0.4524	0.5474	0.7916	−0.6676
	4.0028	−0.0856	−0.1962	1.0000	0.8771	0.3539	0.1525
	4.9992	1.0000	−0.7107	−0.7519	0.7814	−0.0496	0.1994
	7.3734	0.7887	0.5496	0.2616	−0.4802	0.5160	1.0000
4	−1.5757	−0.8290	0.3736	−0.5173	0.5541	−0.2867	1.0000
	1.3851	−0.4672	−0.7019	−0.2773	−0.3067	1.0000	0.1881
	2.6210	0.1112	1.0000	−0.5534	0.5909	0.9014	−0.6366
	4.4961	−0.0953	−0.1684	1.0000	0.8479	0.3498	0.1316
	5.5621	1.0000	−0.8850	−0.8224	0.9321	−0.1301	0.1804
	8.2114	0.8516	0.4719	0.2094	−0.4082	0.4739	1.0000
5	−1.4224	−0.8531	0.3186	−0.4828	0.5029	−0.2478	1.0000
	1.8416	−0.4431	−0.7648	−0.2272	−0.3609	1.0000	0.1854
	3.0988	0.1051	0.9948	−0.6432	0.6273	1.0000	−0.6054
	4.9896	−0.0987	−0.1456	1.0000	0.8274	0.3493	0.1155
	6.1076	0.9227	−0.9783	−0.8281	1.0000	−0.1940	0.1480
	9.0848	0.8953	0.4075	0.1700	−0.3503	0.4351	1.0000

80.

k	λ_i	x_1	x_2	x_3	x_4	x_5	x_6	x_7
0	−5.273	−0.1164	0.0308	0.2179	−0.9777	1.000	0.1220	0.0746
	3.460	0.0490	−0.1719	0.3717	−0.0922	−0.2690	1.000	−0.1757
	4.353	−0.5491	−0.2701	0.6187	−0.2176	−0.4516	−0.2153	1.000
	5.838	−0.1208	1.000	0.1646	−0.1448	−0.2253	0.0386	−0.0230
	10.34	1.000	−0.1052	0.4596	−0.4967	−0.4044	−0.4179	−0.1443
	14.16	0.6454	0.1837	−0.9433	−0.2355	−0.0884	0.4808	1.0000
	19.13	0.6225	0.2802	0.7923	1.000	0.8128	0.1288	0.5397

Continuation of 80

k	λ_i	x_1	x_2	x_3	x_4	x_5	x_6	x_7
1	−6.215	−0.1035	0.0274	0.1889	−0.9690	1.000	0.1065	0.0677
	3.601	0.1192	−0.1196	0.2642	−0.0651	−0.1854	1.000	−0.2733
	4.756	−0.6593	−0.2504	0.6818	−0.2267	−0.4816	0.0378	1.0000
	6.839	−0.0995	1.000	0.1394	−0.1406	−0.2015	0.0370	−0.0406
	11.07	1.000	−0.1667	0.6680	−0.5758	−0.5096	−0.5008	−0.1949
	15.59	0.7518	0.1428	−0.9816	−0.3716	−0.2107	0.3968	1.0000
	20.36	0.7328	0.2950	0.7705	1.000	0.8318	0.1423	0.6536
2	−7.170	−0.0933	0.0248	0.1670	−0.9641	1.000	0.0948	0.0622
	3.690	0.1679	−0.0930	0.2034	−0.0510	−0.1398	1.000	−0.3231
	5.115	−0.7419	−0.2279	0.6832	−0.2185	−0.4705	0.2105	1.000
	7.843	−0.0841	1.000	0.1257	−0.1398	−0.1887	0.0329	−0.0467
	11.80	1.000	−0.2275	0.8792	−0.6608	−0.6176	−0.5614	−0.2275
	17.05	−0.7704	−0.0943	1.000	0.4904	0.3255	−0.3188	−0.9488
	21.67	0.8739	0.3123	0.7425	1.000	0.8493	0.1586	0.7969
3	−8.133	−0.0850	0.0227	0.1498	−0.9614	1.000	0.0856	0.0577
	3.752	0.2033	−0.0769	0.1648	−0.0426	−0.1119	1.000	−0.3518
	5.420	−0.8032	−0.2059	0.6557	−0.2026	−0.4429	0.3330	1.000
	8.847	−0.0725	1.000	0.1181	−0.1406	−0.1814	0.0289	−0.0482
	12.56	0.9139	−0.2641	1.000	−0.6869	−0.6660	−0.5541	−0.2256
	18.48	−0.7251	−0.0462	1.000	0.5840	0.4221	−0.2509	−0.8588
	23.08	1.000	0.3160	0.6748	0.9517	0.8239	0.1688	0.9274

81. $\lambda_1 = 7.9651$, $\lambda_{2,3} = -0.43253 \pm i\ 3.1469$. **82.** $\lambda_1 = 1.5794$, $\lambda_{2,3} = 0.39030 \pm i\ 2.4566$. **83.** $\lambda_1 = -3.1113$, $\lambda_2 = 0.83465$, $\lambda_3 = 3.5053$. $\lambda_4 = 29.771$. **84.** $\lambda_1 = -1.0374$, $\lambda_2 = 0.12705$, $\lambda_3 = 0.54896$, $\lambda_4 = 1.3374$. **85.** $\lambda_1 = -0.806461$, $\lambda_2 = -0.608733$, $\lambda_3 = 0.677051$, $\lambda_4 = 1.05828$;

$$X_1 = \begin{bmatrix} 0.368926 \\ -0.207052 \\ 1.00000 \\ -0.850103 \end{bmatrix}, \quad X_2 = \begin{bmatrix} -0.893682 \\ 1.00000 \\ 0.432658 \\ -0.122454 \end{bmatrix}, \quad X_3 = \begin{bmatrix} 0.135611 \\ -0.094083 \\ 0.780592 \\ 1.00000 \end{bmatrix},$$

$$X_4 = \begin{bmatrix} 1.00000 \\ 0.954806 \\ -0.126324 \\ 0.052826 \end{bmatrix}.$$

86. $\lambda_1 = -0.922196$, $\lambda_2 = -0.701860$, $\lambda_3 = 0.566392$, $\lambda_4 = 1.09388$;

$$X_1 = \begin{bmatrix} -0.027303 \\ 0.006650 \\ 0.898520 \\ 1.00000 \end{bmatrix}, \quad X_2 = \begin{bmatrix} 0.835617 \\ 1.00000 \\ 0.051293 \\ -0.029920 \end{bmatrix}, \quad X_3 = \begin{bmatrix} -0.181129 \\ 0.073013 \\ 1.00000 \\ -0.903951 \end{bmatrix},$$

$$X_4 = \begin{bmatrix} 1.00000 \\ -0.846395 \\ 0.150477 \\ -0.102275 \end{bmatrix}.$$

87. $\lambda_1 = -0.943766$, $\lambda_2 = -0.749386$, $\lambda_3 = 0.680652$, $\lambda_4 = 0.835272$;

$$X_1 = \begin{bmatrix} 0.757683 \\ 1.00000 \\ 0.294756 \\ 0.423991 \end{bmatrix}, \quad X_2 = \begin{bmatrix} -0.294756 \\ -0.423992 \\ 0.757683 \\ 1.00000 \end{bmatrix}, \quad X_3 = \begin{bmatrix} -0.516944 \\ 0.409215 \\ 1.00000 \\ -0.736551 \end{bmatrix},$$

$$X_4 = \begin{bmatrix} 1.00000 \\ -0.736551 \\ 0.516944 \\ -0.409215 \end{bmatrix}.$$

88. $\lambda_1 = -0.958372$, $\lambda_2 = -0.746028$, $\lambda_3 = 0.664545$, $\lambda_4 = 0.835895$;

$$X_1 = \begin{bmatrix} 0.377375 \\ 1.00000 \\ 0.531624 \\ 0.863690 \end{bmatrix}, \quad X_2 = \begin{bmatrix} -0.531624 \\ -0.863690 \\ 0.377375 \\ 1.00000 \end{bmatrix}, \quad X_3 = \begin{bmatrix} -0.640173 \\ 0.188914 \\ 1.00000 \\ -0.554542 \end{bmatrix},$$

$$X_4 = \begin{bmatrix} 1.00000 \\ -0.554543 \\ 0.640173 \\ -0.188914 \end{bmatrix}.$$

89.

k	λ_i	x_1	x_2	x_3
0.0	0.78197	1.0000	−0.32077	−0.77943
	1.0124	−0.20906	1.0000	−0.67977
	2.1056	1.0000	0.84484	0.93529
0.1	0.74934	1.0000	−0.43596	−0.72113
	1.1069	−0.12201	1.0000	−0.77373
	2.3438	1.0000	0.81413	0.89453
0.2	0.71263	1.0000	−0.49518	−0.70037
	1.2045	−0.08114	1.0000	−0.82289
	2.5829	1.0000	0.79407	0.86639

Continuation of 89

k	λ_i	x_1	x_2	x_3
0.3	0.67422	1.0000	−0.53196	−0.69176
	1.3031	−0.05835	1.0000	−0.85335
	2.8226	1.0000	0.78006	0.84573
0.4	0.63492	1.0000	−0.55732	−0.68803
	1.4024	−0.04420	1.0000	−0.87427
	3.0627	1.0000	0.76976	0.82990
0.5	0.59508	1.0000	−0.57601	−0.68652
	1.5018	−0.03474	1.0000	−0.88964
	3.3031	1.0000	0.76191	0.81737

90.

k	λ_i	x_1	x_2	x_3
0	− 35.983	1.0000	−0.29711	−0.61066
	− 16.585	−0.14672	1.0000	−0.72681
	119.17	0.86428	0.85362	1.0000
1	− 35.828	1.0000	−0.27908	−0.62898
	− 15.440	−0.16470	1.0000	−0.70556
	120.87	0.86567	0.84814	1.0000
2	− 35.690	1.0000	−0.26267	−0.64569
	− 14.281	−0.18043	1.0000	−0.68625
	122.57	0.86704	0.84270	1.0000
3	− 35.566	1.0000	−0.24772	−0.66097
	− 13.111	−0.19423	1.0000	−0.66863
	124.28	0.86839	0.83730	1.0000
4	− 35.454	1.0000	−0.23406	−0.67499
	− 11.932	−0.20635	1.0000	−0.65247
	125.99	0.86971	0.83193	1.0000
5	− 35.353	1.0000	−0.22156	−0.68786
	− 10.745	− 0.21702	1.0000	− 0.63759
	127.70	0.87100	0.82661	1.0000
6	− 35.261	1.0000	−0.21008	− 0.69973
	− 9.5513	−0.22643	1.0000	−0.62383
	129.41	0.87228	0.82133	1.0000
7	− 35.177	1.0000	−0.19954	−0.71069
	− 8.3527	−0.23473	1.0000	−0.61105
	131 13	0.87353	0.81609	1.0000

Continuation of 90

k	λ_i	x_1	x_2	x_3
8	− 35.101	1.0000	−0.18981	−0.72083
	− 7.1499	−0.24207	1.0000	−0.59914
	132.85	0.87475	0.81090	1.0000
9	− 35.030	1.0000	−0.18083	−0.73026
	− 5.9437	−0.24856	1.0000	−0.58801
	134.57	0.87596	0.80574	1.0000
10	− 34.966	1.0000	−0.17252	−0.73902
	− 4.7348	−0.25431	1.0000	−0.57756
	136.30	0.87714	0.80062	1.0000
11	− 34.906	1.0000	−0.16481	−0.74720
	− 3.5237	−0.25939	1.0000	−0.56772
	138.03	0.87831	0.79555	1.0000
12	− 34.851	1.0000	−0.15764	−0.75484
	− 2.3111	−0.26389	1.0000	−0.55844
	139.76	0.87945	0.79051	1.0000
13	− 34.799	1.0000	−0.15097	−0.76199
	− 1.0972	−0.26786	1.0000	−0.54965
	141.50	0.88058	0.78552	1.0000

91.

k	λ_i	x_1	x_2	x_3	x_4
0	0.027661	1.0000	−0.41317	−0.26660	−0.19892
	1.6933	0.19039	1.0000	−0.63541	−0.26833
	2.6035	−0.15467	−0.27459	−0.90079	1.0000
	7.6755	0.67811	0.64559	0.79690	1.0000
1	0.14444	1.0000	−0.40793	−0.25558	−0.20466
	1.7074	0.18598	1.0000	0.67897	−0.23654
	2.7312	−0.14352	−0.31758	−0.85542	1.0000
	7.8169	0.68079	0.65979	0.80985	1.0000
2	0.26041	1.0000	−0.40357	−0.24443	−0.21019
	1.7186	0.18430	1.0000	−0.71532	−0.21141
	2.8618	−0.13132	−0.35021	−0.81896	1.0000
	7.9592	0.68349	0.67424	0.82313	1.0000

Continuation of 91

k	λ_i	x_1	x_2	x_3	x_4
3	0.37559	1.0000	−0.40023	−0.23302	−0.21550
	1.7277	0.18506	1.0000	−0.74629	−0.19149
	2.9943	−0.11852	−0.37530	−0.78887	1.0000
	8.1023	0.68623	0.68896	0.83676	1.0000
4	0.49000	1.0000	−0.39805	−0.22119	−0.22060
	1.7354	0.18823	1.0000	−0.77332	−0.17574
	3.1283	−0.10541	−0.39479	−0.76342	1.0000
	8.2464	0.68899	0.70395	0.85074	1.0000
5	0.60363	1.0000	0.39727	−0.20874	−0.22548
	1.7420	0.19394	1.0000	−0.79754	−0.16345
	3.2631	−0.092161	−0.40999	−0.74142	1.0000
	8.3913	0.69178	0.71921	0.86506	1.0000

92.

k	λ_i	x_1	x_2	x_3	x_4
0.0	− 1.5696	−0.32703	−0.40171	0.10054	1.0000
	0.80127	1.0000	−0.90232	−0.93942	0.05901
	4.4386	−0.048571	1.0000	−0.98178	0.48454
	15.690	1.0000	0.46516	0.64589	0.44895
0.1	− 1.5784	−0.35201	−0.38513	0.12070	1.0000
	0.78832	1.0000	−0.93074	−0.95626	0.10898
	4.4908	−0.050652	1.0000	−0.97107	0.48450
	15.859	1.0000	0.46134	0.64816	0.45145
0.2	− 1.5922	−0.37709	−0.36855	0.14119	1.0000
	0.77991	1.0000	−0.96076	−0.97368	0.16047
	4.5431	−0.052529	1.0000	−0.96085	0.48440
	16.029	1.0000	0.45755	0.65035	0.45390
0.3	− 1.6110	−0.40213	−0.35209	0.16188	1.0000
	0.77602	1.0000	−0.99224	−0.99168	0.21330
	4.5954	−0.054222	1.0000	−0.95109	0.48425
	16.200	1.0000	0.45381	0.65247	0.45630
0.4	− 1.6346	−0.42701	−0.33586	0.18263	1.0000
	0.77656	−0.97561	1.0000	0.98560	−0.26074
	4.6478	−0.055749	1.0000	−0.94174	0.48404
	16.370	1.0000	0.45011	0.65450	0.45865

Continuation of 92

k	λ_i	x_1	x_2	x_3	x_4
0.5	− 1.6630	−0.45159	−0.31993	0.20332	1.0000
	0.78140	−0.94440	1.0000	0.97212	−0.30421
	4.7002	−0.057126	1.0000	−0.93278	0.48379
	16.541	1.0000	0.44646	0.65646	0.46095
0.6	− 1.6959	−0.47578	−0.30441	0.22384	1.0000
	0.79036	−0.91434	1.0000	0.95913	−0.34530
	4.7527	−0.058367	1.0000	−0.92418	0.48351
	16.713	1.0000	0.44284	0.65836	0.46321
0.7	− 1.7330	−0.49946	−0.28935	0.24407	1.0000
	0.80324	−0.88555	1.0000	0.94673	−0.38401
	4.8051	−0.059484	1.0000	−0.91591	0.48319
	16.885	1.0000	0.43926	0.66018	0.46543
0.8	− 1.7742	−0.52256	−0.27482	0.26392	1.0000
	0.81980	−0.85809	1.0000	0.93500	−0.42036
	4.8576	−0.060488	1.0000	−0.90796	0.48284
	17.057	1.0000	0.43573	0.66195	0.46761
0.9	− 1.8192	−0.54502	−0.26085	0.28333	1.0000
	0.83978	−0.83198	1.0000	0.92397	−0.45439
	4.9101	−0.061390	1.0000	−0.90030	0.48247
	17.229	1.0000	0.43224	0.66365	0.46974
1.0	− 1.8677	−0.56679	−0.24747	0.30222	1.0000
	0.86293	−0.80724	1.0000	0.91367	−0.48619
	4.9626	−0.062199	1.0000	−0.89292	0.48208
	17.402	1.0000	0.42879	0.66529	0.47184
1.1	− 1.9194	−0.58783	−0.23470	0.32056	1.0000
	0.88900	−0.78383	1.0000	0.90410	−0.51587
	5.0152	−0.062923	1.0000	−0.88579	0.48166
	17.575	1.0000	0.42538	0.66688	0.47389
1.2	− 1.9742	−0.60813	−0.22255	0.33831	1.0000
	0.91774	−0.76172	1.0000	0.89526	−0.54355
	5.0677	−0.063568	1.0000	−0.87890	0.48124
	17.749	1.0000	0.42201	0.66841	0.47591
1.3	− 2.0316	−0.62767	−0.21101	0.35546	1.0000
	0.94890	−0.74086	1.0000	0.88712	−0.56934
	5.1202	−0.064142	1.0000	−0.87224	0.48080
	17.923	1.0000	0.41868	0.66989	0.47790

Continuation of 92

k	λ_i	x_1	x_2	x_3	x_4
1.4	− 2.0916	−0.64647	−0.20006	0.37199	1.0000
	0.98226	−0.72118	1.0000	0.87964	−0.59338
	5.1727	−0.064651	1.0000	−0.86580	0.48034
	18.097	1.0000	0.41539	0.67132	0.47985
1.5	− 2.1538	−0.66453	−0.18971	0.38791	1.0000
	1.0176	−0.70262	1.0000	0.87281	−0.61578
	5.2252	−0.065099	1.0000	−0.85956	0.47988
	18.271	1.0000	0.41214	0.67270	0.48176

93.

k	λ_i	x_1	x_2	x_3	x_4	x_5
0	−4.8635	0.092010	1.0000	0.024360	−0.78876	−0.26244
	−0.44548	0.071723	−0.34837	1.0000	−0.73438	0.99770
	0.14576	−0.43756	0.24646	−0.95448	−0.10079	1.0000
	1.9504	1.0000	−0.11757	−0.32574	−0.091187	0.14643
	11.813	0.28965	0.88185	0.51768	1.0000	0.50430
1	−4.8264	0.11286	1.0000	0.003467	−0.81966	−0.23290
	−0.20371	−0.05275	−0.32007	1.0000	−0.54167	0.52141
	0.43180	−0.70263	0.21330	−0.55683	−0.12301	1.0000
	3.6628	1.0000	−0.14227	−0.35818	−0.18260	0.51107
	14.536	0.28587	0.89407	0.59944	1.0000	0.46694
2	−4.8043	0.12444	1.0000	−0.010189	−0.83860	−0.21603
	−0.096557	−0.10933	−0.32279	1.0000	−0.52535	0.43502
	0.54298	−0.83055	0.23791	−0.49689	−0.091104	1.0000
	5.6350	1.0000	−0.15346	−0.34290	−0.20510	0.67800
	17.323	0.27637	0.90446	0.65916	1.0000	0.43299
3	−4.7895	0.13166	1.0000	−0.019894	−0.85143	−0.20523
	−0.029758	−0.15292	−0.32721	1.0000	−0.53024	0.41039
	0.60090	−0.89419	0.26281	−0.49249	−0.059142	1.0000
	7.6647	1.0000	−0.15539	−0.32049	−0.20480	0.76507
	20.154	0.26448	0.91322	0.70431	1.0000	0.40250

Continuation of 93

k	λ_i	x_1	x_2	x_3	x_4	x_5
4	−4.7790	0.13654	1.0000	−0.027177	−0.86072	−0.19778
	0.016842	−0.18869	−0.33096	1.0000	−0.53890	0.40419
	0.63775	−0.93004	0.28426	−0.50241	−0.031205	1.0000
	9.7089	1.0000	−0.15276	−0.29852	−0.19723	0.81738
	23.015	0.25189	0.92065	0.73945	1.0000	0.37528
5	−4.7710	0.14003	1.0000	−0.032858	−0.86777	−0.19235
	0.051336	−0.21865	−0.33392	1.0000	−0.54773	0.40498
	0.66395	−0.95211	0.30242	−0.51615	−0.007255	1.0000
	11.755	1.0000	−0.14803	−0.27844	−0.18748	0.85180
	25.900	0.23941	0.92697	0.76747	1.0000	0.35099
6	−4.7648	0.14264	1.0000	−0.037421	−0.87331	−0.18823
	0.077898	−0.24407	−0.33624	1.0000	−0.55581	0.40867
	0.68387	−0.96656	0.31785	−0.53033	0.013283	1.0000
	13.800	1.0000	−0.14240	−0.26042	−0.17738	0.87607
	28.803	0.22747	0.93242	0.79027	1.0000	0.32931
7	−4.7598	0.14465	1.0000	−0.041169	0.87777	−0.18499
	0.098965	−0.26588	−0.33806	1.0000	−0.56300	0.41352
	0.69969	−0.97646	0.33106	−0.54378	0.030990	1.0000
	15.842	1.0000	−0.13649	−0.24433	−0.16765	0.89398
	31.719	0.21622	0.93714	0.80915	1.0000	0.30989
8	−4.7557	0.14625	1.0000	−0.04430	−0.88145	−0.18240
	0.11607	−0.28477	−0.33951	1.0000	−0.56933	0.41874
	0.71264	−0.98348	0.34246	−0.55613	0.04637	1.0000
	17.881	1.0000	−0.13063	−0.22994	−0.15855	0.90769
	34.646	0.20574	0.94126	0.82502	1.0000	0.29246
9	−4.7523	0.14754	1.0000	−0.046968	−0.88454	−0.18027
	0.13021	−0.30125	−0.34068	1.0000	−0.57490	0.42395
	0.72349	−0.98859	0.35238	−0.56734	0.05980	1.0000
	19.916	1.0000	−0.12497	−0.21703	−0.15014	0.91848
	37.582	0.19602	0.94489	0.83853	1.0000	0.27675
10	−4.7494	0.14861	1.0000	−0.04926	−0.88717	−0.17849
	0.14210	−0.31576	−0.34164	1.0000	−0.57981	0.42899
	0.73272	−0.99240	0.36108	−0.57745	0.07164	1.0000
	21.948	1.0000	−0.11959	−0.20542	−0.14242	0.92716
	40.526	0.18703	0.94811	0.85015	1.0000	0.26254

94. k	λ_i	x_1	x_2	x_3	x_4	x_5	x_6
0	−10.27	0.8702	1.000	−0.4963	−0.4274	−0.4275	0.8545
	− 4.122	0.02964	−0.1200	−0.9363	0.01537	1.000	0.07450
	1.420	1.000	−0.00811	0.8577	−0.9591	0.8300	−0.5752
	3.396	−0.8155	−0.2599	0.3430	−0.9216	0.2538	1.000
	11.43	−0.4704	1.000	0.3607	0.2929	0.4744	−0.0979
	21.74	0.5758	−0.3767	0.6448	1.000	0.4548	0.9567
1	−9.859	0.8639	1.000	−0.4803	−0.4232	−0.4756	0.8465
	−3.942	0.03155	−0.1089	−0.9679	0.00357	1.000	0.1110
	1.646	1.000	−0.00917	0.7850	−0.9093	0.7941	−0.5727
	3.434	−0.7966	−0.2513	0.3824	−0.9460	0.2602	1.000
	11.57	−0.4576	1.000	0.3509	0.2742	0.4744	−0.1115
	21.96	0.5768	−0.3498	0.6468	1.000	0.4606	0.9503
2	−9.466	0.8575	1.000	−0.4616	−0.4182	−0.5254	0.8367
	−3.758	−0.03375	0.09610	1.000	0.00893	−0.9969	−0.1501
	1.870	1.000	−0.00969	0.7099	−0.8533	0.7560	−0.5733
	3.477	−0.7732	−0.2439	0.4267	−0.9798	0.2714	1.000
	11.70	−0.4453	1.000	0.3412	0.2553	0.4749	−0.1248
	22.17	0.5782	−0.3231	0.6488	1.000	0.4669	0.9445
3	−9.089	0.8513	1.000	−0.4403	−0.4123	−0.5768	0.8250
	−3.571	−0.03507	−0.07891	1.000	0.02137	−0.9591	−0.1856
	2.089	1.000	−0.00947	0.6315	−0.7899	0.7153	−0.5779
	3.527	0.7244	0.2321	−0.4653	1.000	−0.2817	−0.9743
	11.84	−0.4333	1.000	0.3315	0.2360	0.4757	−0.1376
	22.40	0.5800	−0.2965	0.6509	1.000	0.4736	0.9391
4	−8.728	0.8452	1.000	−0.4164	−0.4055	−0.6294	0.8111
	−3.380	−0.03648	0.06104	1.000	0.03356	−0.9200	−0.2211
	2.304	1.000	−0.00834	0.5494	−0.7178	0.6718	−0.5873
	3.585	0.6464	0.2154	−0.4927	1.000	−0.2900	−0.9173
	11.98	−0.4217	1.000	0.3217	0.2163	0.4768	−0.1501
	22.64	0.5824	−0.2700	0.6531	1.000	0.4809	0.9343
5	−8.385	0.8396	1.000	−0.3898	−0.3978	−0.6830	0.7949
	−3.184	−0.03788	0.04261	1.000	0.04542	−0.8797	−0.2564
	2.512	1.000	−0.00613	0.4629	−0.6359	0.6255	−0.6021
	3.654	0.5543	0.1993	−0.5167	1.000	−0.3025	−0.8489
	12.12	−0.4104	1.000	0.3121	0.1964	0.4783	−0.1623
	22.88	0.5852	−0.2436	0.6554	1.000	0.4886	0.9299

95.

k	λ_i	x_1	x_2	x_3	x_4	x_5	x_6	x_7
0	−4.026	−0.3354	0.1078	−0.5129	0.3735	1.000	−0.02914	−0.8577
	−1.967	−0.4418	0.1336	0.6190	−0.2160	−0.1999	1.000	−0.5417
	−0.5367	−0.2489	−0.4028	0.3477	−0.7716	1.000	0.1218	0.6646
	0.1335	−0.1967	1.000	−0.09852	−0.5293	−0.008603	−0.2594	0.02982
	1.378	1.000	−0.002393	−0.5820	−0.5663	0.08354	0.5818	−0.2123
	5.405	0.9346	0.6756	1.000	0.7139	0.6083	0.05516	0.1396
	6.813	−0.3567	0.4412	−0.7095	0.6955	0.09400	0.9334	1.000
1	−2.891	−0.3317	0.1409	−0.5939	0.3347	1.000	−0.03969	−0.8987
	−1.207	−0.5167	0.1042	0.7585	−0.6283	0.2202	1.000	−0.3283
	−0.3769	0.008865	−0.2055	0.03399	−0.7448	1.000	−0.4265	0.7960
	1.236	−0.2055	1.000	−0.03691	−0.3006	−0.1528	−0.3411	−0.01049
	2.637	1.000	0.2101	−0.4692	−0.4833	0.03984	0.4782	−0.1829
	7.132	0.7134	0.2709	1.000	0.4424	0.4093	−0.3115	−0.2477
	7.669	−0.03473	0.6341	−0.1964	0.9248	0.4068	0.8839	1.000
2	−1.795	−0.3250	0.1710	−0.6984	0.3470	1.000	−0.06674	−0.9592
	−1.072	0.4326	−0.04940	−0.7145	1.000	−0.7607	−0.6441	−0.02168
	0.3003	−0.3184	0.2317	0.3750	0.3919	−0.8529	1.000	−0.9408
	2.307	−0.2432	1.000	0.01966	−0.2163	−0.1822	−0.3827	0.00480
	3.843	1.000	0.3256	−0.3742	−0.4046	0.02304	0.4116	−0.1593
	8.261	0.6191	−0.2443	1.000	−0.1542	0.03105	−0.9194	−0.9408
	9.355	0.4217	0.5288	0.6844	1.000	0.6143	0.2961	0.4347
3	−1.259	0.2607	−0.00944	−0.7052	1.000	−0.6434	−0.4145	−0.2304
	−0.7151	−0.3359	0.1767	−0.5672	0.1377	1.000	0.00592	−0.8570
	1.233	−0.3567	0.2092	0.4338	0.2037	−0.6542	1.000	−0.8279
	3.351	−0.2729	1.000	0.05842	−0.1695	−0.1946	−0.4103	0.01730
	4.995	1.000	0.3893	−0.3028	−0.3408	0.01652	0.3675	−0.1442
	9.148	−0.5238	0.3407	−0.8552	0.2281	0.09877	0.9589	1.000
	11.45	0.3931	0.4136	0.7938	1.000	0.6262	0.1623	0.2983

96.

k	Characteristic equations
0.0	$(\lambda - 0.9)(\lambda^3 - 1.5\lambda^2 + 0.38\lambda - 0.84) = 0$
0.1	$(\lambda - 1)(\lambda^3 - 1.62\lambda^2 + 0.548\lambda - 1.08) = 0$
0.2	$(\lambda - 1.1)(\lambda^3 - 1.74\lambda^2 + 0.736\lambda - 1.34) = 0$
0.3	$(\lambda - 1.2)(\lambda^3 - 1.86\lambda^2 + 0.944\lambda - 1.62) = 0$
0.4	$(\lambda - 1.3)(\lambda^3 - 1.98\lambda^2 + 1.172\lambda - 1.92) = 0$
0.5	$(\lambda - 1.4)(\lambda^3 - 2.1\lambda^2 + 1.42\lambda - 2.24) = 0$
0.6	$(\lambda - 1.5)(\lambda^3 - 2.22\lambda^2 + 1.688\lambda - 2.58) = 0$
0.7	$(\lambda - 1.6)(\lambda^3 - 2.34\lambda^2 + 1.976\lambda - 2.94) = 0$
0.8	$(\lambda - 1.7)(\lambda^3 - 2.46\lambda^2 + 2.284\lambda - 3.32) = 0$
0.9	$(\lambda - 1.8)(\lambda^3 - 2.58\lambda^2 + 2.612\lambda - 3.72) = 0$
1.0	$(\lambda - 1.9)(\lambda^3 - 2.7\lambda^2 + 2.96\lambda - 4.14) = 0$

97. $p_1 = 3.86$, $p_2 = -10.78$, $p_3 = 15.172$. **98.** $p_1 = -0.166176$, $p_2 = 1.27966$, $p_3 = 0.100320$, $p_4 = -0.396302$. **99.** $p_1 = -0.209302$, $p_2 = 1.27774$, $p_3 = 0.135842$, $p_4 = -0.396299$.

100. $\lambda_1 = -0.930012$, $\lambda_2 = -0.754496$, $\lambda_3 = 0.656227$, $\lambda_4 = 0.867255$;

$$X_1 = \begin{bmatrix} -0.590771 \\ 0.551160 \\ 0.390967 \\ 1.00000 \end{bmatrix}, X_2 = \begin{bmatrix} 0.390967 \\ 1.00000 \\ 0.590771 \\ -0.551160 \end{bmatrix}, X_3 = \begin{bmatrix} 1.00000 \\ 0.327855 \\ -0.611230 \\ 0.649041 \end{bmatrix}, X_4 = \begin{bmatrix} 0.611230 \\ -0.649041 \\ 1.00000 \\ 0.327855 \end{bmatrix};$$

$$X^{-1} = \begin{bmatrix} 0.707460 & -0.070913 & -0.151135 & 1.15135 \\ -0.070913 & -0.573487 & -1.11589 & 0.117829 \\ -0.151135 & -1.11589 & 0.558192 & -0.112877 \\ 1.15135 & 0.117829 & -0.112877 & -0.415881 \end{bmatrix}.$$

01. $\lambda_1 = -0.942979$, $\lambda_2 = -0.769905$, $\lambda_3 = 0.688271$, $\lambda_4 = 0.852096$;

$$X_1 = \begin{bmatrix} 1.00000 \\ -0.493957 \\ -0.150160 \\ -0.946949 \end{bmatrix}, X_2 = \begin{bmatrix} 0.150160 \\ 0.946949 \\ 1.00000 \\ -0.493957 \end{bmatrix}, X_3 = \begin{bmatrix} 1.00000 \\ 0.460067 \\ -0.169459 \\ 0.842910 \end{bmatrix}, X_4 = \begin{bmatrix} 0.169459 \\ -0.842910 \\ 1.00000 \\ 0.460067 \end{bmatrix};$$

$$X^{-1} = \begin{bmatrix} 0.258267 & 0.413480 & -0.040812 & 1.18341 \\ 0.413480 & -0.072970 & -1.17006 & 0.107074 \\ -0.040812 & -1.17006 & 0.011477 & 0.397256 \\ 1.18341 & 0.107074 & 0.397256 & 0.070388 \end{bmatrix}.$$

02. $\lambda_1 = -0.959926$, $\lambda_2 = -0.749037$, $\lambda_3 = 0.689166$, $\lambda_4 = 0.852762$;

$$X_1 = \begin{bmatrix} -0.787236 \\ 0.516755 \\ 0.304536 \\ 1.00000 \end{bmatrix}, X_2 = \begin{bmatrix} 0.304536 \\ 1.00000 \\ 0.787236 \\ -0.516755 \end{bmatrix}, X_3 = \begin{bmatrix} 1.00000 \\ 0.395423 \\ -0.422126 \\ 0.711452 \end{bmatrix}, X_4 = \begin{bmatrix} 0.422126 \\ -0.711452 \\ 1.00000 \\ 0.395423 \end{bmatrix};$$

$$X^{-1} = \begin{bmatrix} 0.513121 & 0.129085 & -0.099360 & 1.18760 \\ 0.129085 & -0.369244 & -1.19858 & 0.119111 \\ -0.099360 & -1.19858 & 0.310756 & 0.129266 \\ 1.18760 & 0.119111 & 0.129266 & -0.207741 \end{bmatrix}.$$

103. $\lambda_1 = -0.932102$, $\lambda_2 = -0.757360$, $\lambda_3 = 0.683667$, $\lambda_4 = 0.868596$;

$$X_1 = \begin{bmatrix} 1.00000 \\ 0.080751 \\ -0.489793 \\ -0.980059 \end{bmatrix}, \quad X_2 = \begin{bmatrix} 0.489793 \\ 0.980059 \\ 1.00000 \\ 0.080752 \end{bmatrix}, \quad X_3 = \begin{bmatrix} 0.762437 \\ -0.008624 \\ -0.445736 \\ 1.00000 \end{bmatrix}, \quad X_4 = \begin{bmatrix} -0.445736 \\ 1.00000 \\ -0.762437 \\ 0.008624 \end{bmatrix}$$

$$A^{-1} = \begin{bmatrix} -0.023483 & -0.620137 & -0.114391 & 1.07678 \\ -0.620137 & 0.068993 & -1.05709 & -0.010386 \\ -0.114391 & -1.05709 & -0.175676 & -0.652184 \\ 1.07678 & -0.010386 & -0.652184 & 0.350928 \end{bmatrix}$$

104.

k	λ_i	x_1	x_2	x_3
0	0.035327	1.0000	−0.90891	0.10514
	1.3318	−0.42216	−0.34880	1.0000
	2.8538	0.83517	1.0000	0.70137
1	0.042248	1.0000	−0.94187	0.072315
	2.4757	−0.25251	−0.19131	1.0000
	4.7030	0.91139	1.0000	0.42145
2	0.045636	1.0000	−0.95704	0.05473
	3.5371	−0.17591	−0.12662	1.0000
	6.6382	0.94105	1.0000	0.29216
3	0.047646	1.0000	−0.96589	0.043970
	4.5694	−0.13423	−0.093448	1.0000
	8.6039	0.95614	1.0000	0.22179
4	0.048976	1.0000	−0.97171	0.036727
	5.5890	−0.10833	−0.07369	1.0000
	10.583	0.96516	1.0000	0.17824
5	0.049920	1.0000	−0.97582	0.031527
	6.6021	−0.090742	−0.060682	1.0000
	12.569	0.97113	1.0000	0.14880
6	0.050626	1.0000	−0.97889	0.027613
	7.6114	−0.078043	−0.051517	1.0000
	14.559	0.97537	1.0000	0.12764
7	0.051174	1.0000	−0.98127	0.024563
	8.6184	−0.068450	−0.044725	1.0000
	16.551	0.97853	1.0000	0.11171
8	0.051611	1.0000	−0.98317	0.022119
	9.6238	−0.060952	−0.039499	1.0000
	18.546	0.98097	1.0000	0.099291

Continuation of 104

k	λ_i	x_1	x_2	x_3
9	0.051967	1.0000	−0.98471	0.020116
	10.628	−0.054931	−0.035355	1.0000
	20.541	0.98292	1.0000	0.089348
10	0.052264	1.0000	−0.98600	0.018446
	11.632	−0.04999	−0.031993	1.0000
	22.537	0.98450	1.0000	0.081209
11	0.052515	1.0000	−0.98709	0.017032
	12.635	−0.045866	−0.029211	1.0000
	24.534	0.98582	1.0000	0.074425
12	0.052730	1.0000	−0.98802	0.015819
	13.637	−0.042368	−0.026871	1.0000
	26.531	0.98693	1.0000	0.068685
13	0.052916	1.0000	−0.98882	0.014767
	14.639	−0.03937	−0.024876	1.0000
	28.529	0.98788	1.0000	0.063764
14	0.053078	1.0000	−0.98952	0.013847
	15.641	−0.036760	−0.023156	1.0000
	30.527	0.98870	1.0000	0.059500
15	0.05322	1.0000	−0.99014	0.01303
	16.642	−0.034478	−0.021657	1.0000
	32.525	0.98942	1.0000	0.055770

105.

k	λ_i	x_1	x_2	x_3	x_4
0	−4.8660	−0.94294	0.15233	0.77702	1.0000
	−0.56603	−0.42564	1.0000	−0.26867	−0.34492
	−0.046067	−0.11858	−0.083946	1.0000	−0.87604
	23.078	1.0000	0.71374	0.54103	0.41383
1	−4.7470	−0.92194	0.15571	0.71355	1.0000
	0.007058	−0.37826	1.0000	−0.48873	−0.15571
	0.54780	−0.26925	0.23151	1.0000	−0.99783
	23.792	1.0000	0.70668	0.53492	0.43021
2	−4.6390	−0.90664	0.15888	0.66106	1.0000
	0.52767	−0.33596	1.0000	−0.67003	−0.02054
	1.1967	0.35974	−0.46647	−0.90723	1.0000
	24.515	1.0000	0.69932	0.52864	0.44607

Continuation of 105

k	λ_i	x_1	x_2	x_3	x_4
3	−4.5403	−0.89531	0.16158	0.61684	1.0000
	1.0234	−0.30254	1.0000	−0.79815	0.05989
	1.8714	0.41445	−0.61891	−0.85749	1.0000
	25.246	1.0000	0.69170	0.52220	0.46144
4	−4.4495	−0.88683	0.16372	0.57901	1.0000
	1.5099	−0.27609	1.0000	−0.88871	0.10601
	2.5551	0.44904	−0.72380	−0.83465	1.0000
	25.984	1.0000	0.68385	0.51564	0.47630
5	−4.3655	−0.88043	0.16533	0.54623	1.0000
	1.9935	−0.25452	1.0000	−0.95544	0.13248
	3.2406	0.47224	−0.80220	−0.82675	1.0000
	26.731	1.0000	0.67580	0.50897	0.49069
6	−4.2874	−0.87562	0.16644	0.51752	1.0000
	2.4766	0.23486	−0.99333	1.0000	−0.14654
	3.9248	0.48852	−0.86509	−0.82751	1.0000
	27.486	1.0000	0.66759	0.50222	0.50459
7	−4.2145	−0.87201	0.16711	0.49213	1.0000
	2.9603	0.21095	−0.95466	1.0000	−0.14864
	4.6060	0.50034	−0.91833	−0.83360	1.0000
	28.248	1.0000	0.65925	0.49541	0.51804
8	−4.1461	−0.86934	0.16741	0.46948	1.0000
	3.4450	0.19200	−0.92523	1.0000	−0.14768
	5.2834	0.50908	−0.96526	−0.84315	1.0000
	29.018	1.0000	0.65079	0.48855	0.53103
9	−4.0818	−0.86741	0.16738	0.44914	1.0000
	3.9307	0.17650	−0.90200	1.0000	−0.14507
	5.9568	−0.51158	1.0000	0.84836	−0.99216
	29.794	1.0000	0.64226	0.48167	0.54357
10	−4.0212	−0.86607	0.16707	0.43076	1.0000
	4.4175	0.16353	−0.88316	1.0000	−0.14158
	6.6257	−0.49690	1.0000	0.82925	−0.95463
	30.578	1.0000	0.63367	0.47478	0.55569

106.

k	λ_i	x_1	x_2	x_3	x_4
0	−0.71696	−0.32094	1.0000	0.009388	−0.071321
	−0.35573	−0.75037	−0.23597	1.0000	0.19965
	5.0273	−0.090371	0.044728	−0.25690	1.0000
	10.445	1.0000	0.33324	0.77423	0.27437
1	−0.42403	0.18800	1.0000	−0.67723	−0.15720
	1.0459	1.0000	−0.75934	−0.81642	−0.11724
	6.0155	−0.075223	0.007178	−0.24241	1.0000
	11.763	1.0000	0.40348	0.81097	0.26892
2	−0.33610	0.21692	1.0000	−0.76502	−0.14259
	2.5797	1.0000	−0.76923	−0.69785	−0.12933
	7.0194	−0.051191	−0.023982	−0.23225	1.0000
	13.137	1.0000	0.47716	0.85844	0.26201
3	−0.27042	0.21279	1.0000	−0.80365	−0.12608
	4.0663	1.0000	−0.72289	−0.61056	−0.15412
	8.0356	−0.01989	−0.05060	−0.22511	1.0000
	14.568	1.0000	0.55544	0.91604	0.25421
4	−0.21759	0.20247	1.0000	−0.82786	−0.11194
	5.4990	1.0000	−0.66764	−0.53688	−0.18492
	9.0620	0.01751	−0.073781	−0.22005	1.0000
	16.057	1.0000	0.63903	0.98322	0.24599
5	−0.17385	0.19111	1.0000	−0.84539	−0.10019
	6.8775	1.0000	−0.61310	−0.47307	−0.22018
	10.097	0.060086	−0.09419	−0.21634	1.0000
	17.600	0.94396	0.68747	1.0000	0.22438
6	−0.13697	0.18011	1.0000	−0.85905	−0.090427
	8.2039	1.0000	−0.56227	−0.41760	−0.25904
	11.138	0.10685	−0.11216	−0.21343	1.0000
	19.195	0.87428	0.71973	1.0000	0.20074
7	−0.10542	0.16989	1.0000	−0.87016	−0.082247
	9.4820	1.0000	−0.51611	−0.36946	−0.30065
	12.186	0.15676	−0.12786	−0.21085	1.0000
	20.838	0.80922	0.74746	1.0000	0.17957
8	−0.078113	0.16053	1.0000	−0.87947	−0.075325
	10.716	1.0000	−0.47477	−0.32783	−0.34410
	13.237	0.20871	−0.14135	−0.20827	1.0000
	22.524	0.74938	0.77129	1.0000	0.16089

Continuation of 106

k	λ_i	x_1	x_2	x_3	x_4
9	−0.054247	0.15202	1.0000	−0.88742	−0.069408
	11.913	1.0000	−0.43804	−0.29192	−0.38843
	14.292	0.26157	−0.15268	−0.20545	1.0000
	24.250	0.69492	0.79182	1.0000	0.14457
10	−0.033207	0.14428	1.0000	−0.89433	−0.064304
	13.076	1.0000	−0.40551	−0.26099	−0.43270
	15.348	0.31426	−0.16190	−0.20224	1.0000
	26.009	0.64569	0.80955	1.0000	0.13039

107.

k	λ_i	x_1	x_2	x_3	x_4	x_5
0	−10.173	0.10833	−0.51509	−0.27579	−0.39446	1.0000
	− 9.2617	−0.60276	−0.47626	0.76550	1.0000	0.42556
	0.24342	1.0000	−0.64970	−0.067652	0.46374	−0.27872
	3.7220	0.65928	0.55468	1.0000	−0.26758	0.38452
	15.469	0.32757	1.0000	−0.76106	0.97744	0.65527
1	− 9.9019	0.35495	−0.49455	−0.51701	−0.67620	1.0000
	− 8.8494	−0.68333	−0.75294	0.83641	1.0000	0.97881
	− 0.11131	1.0000	−0.76707	−0.13983	0.60910	−0.39474
	5.3353	0.78028	0.47950	1.0000	−0.24627	0.31066
	16.027	0.34314	1.0000	−0.71394	0.95135	0.64693
2	− 9.8398	0.63096	−0.52494	−0.77427	−0.92629	1.0000
	− 8.3441	−0.47050	−0.67133	0.59088	0.64563	1.0000
	− 0.43010	1.0000	−0.84358	−0.14980	0.75225	−0.49298
	7.0102	0.82485	0.43174	1.0000	−0.25028	0.24863
	16.604	0.36504	1.0000	−0.66052	0.92255	0.63774
3	− 9.9359	−0.82476	0.55182	0.91589	1.0000	−0.90600
	− 7.8016	−0.36818	−0.62985	0.49340	0.49800	1.0000
	− 0.66578	1.0000	−0.90788	−0.12905	0.90761	−0.59196
	8.6978	0.83547	0.38651	1.0000	−0.26729	0.19076
	17.205	0.39458	1.0000	−0.59804	0.88959	0.62719
4	−10.162	−0.96987	0.59681	1.0000	0.99279	−0.80975
	− 7.2498	−0.30945	−0.60656	0.44983	0.42486	1.0000
	− 0.80362	0.92571	−0.89800	−0.082222	1.0000	−0.64610
	10.372	0.82419	0.33433	1.0000	−0.29449	0.13313
	17.844	0.43366	1.0000	−0.52306	0.85067	0.61462
5	−10.504	1.0000	−0.60013	−0.97182	−0.87511	0.66657
	− 6.6982	−0.27025	−0.59114	0.42815	0.38280	1.0000
	− 0.84270	0.78668	−0.81356	−0.027206	1.0000	−0.63948
	12.010	0.79442	0.26942	1.0000	−0.33196	0.072879
	18.536	0.48476	1.0000	−0.43141	0.80366	0.59922

Continuation of 107

k	λ_i	x_1	x_2	x_3	x_4	x_5
6	−10.946	1.0000	−0.59993	−0.92629	−0.75627	0.54916
	− 6.1498	−0.24157	−0.57992	0.41688	0.35617	1.0000
	− 0.79047	0.67611	−0.74481	0.020404	1.0000	−0.63328
	13.578	0.74662	0.18747	1.0000	−0.38064	0.007776
	19.309	0.55052	1.0000	−0.31890	0.74641	0.58000
7	−11.474	1.0000	−0.60555	−0.89087	−0.66052	0.46296
	− 5.6053	−0.21929	−0.57125	0.41108	0.33817	1.0000
	− 0.65887	0.58770	−0.68918	0.060439	1.0000	−0.62783
	15.035	0.68146	0.086732	1.0000	−0.44074	−0.063046
	20.203	0.63220	1.0000	−0.18370	0.67791	0.55615
8	−12.071	1.0000	−0.61309	−0.86306	−0.58289	0.39749
	− 5.0649	−0.20130	−0.56429	0.40838	0.32544	1.0000
	− 0.46118	0.51654	−0.64414	0.093676	1.0000	−0.62320
	16.337	0.60297	−0.028978	1.0000	−0.50951	−0.13754
	21.260	0.72681	1.0000	−0.030735	0.60048	0.52773
9	−12.725	1.0000	−0.62088	−0.84101	−0.51947	0.34649
	− 4.5283	−0.18635	−0.55854	0.40751	0.31612	1.0000
	− 0.20987	0.45877	−0.60751	0.12117	1.0000	−0.61933
	17.457	0.52003	−0.14877	1.0000	−0.58033	−0.21024
	22.505	0.82554	1.0000	0.12610	0.52089	0.49633
10	−13.423	1.0000	−0.62824	−0.82335	−0.46716	0.30594
	− 3.9950	−0.17365	−0.55370	0.40776	0.30912	1.0000
	0.084330	0.41141	−0.57747	0.14396	1.0000	−0.61613
	18.405	0.44266	−0.25983	1.0000	−0.64566	−0.27517
	23.928	0.91760	1.0000	0.27028	0.44729	0.46457

108.

k	λ_i	x_1	x_2	x_3	x_4	x_5	x_6
0	− 7.598	0.4813	−0.4349	−0.3919	0.9045	−0.3215	1.000
	− 2.964	0.6042	1.000	0.03379	−0.09497	−0.7211	0.01146
	− 0.6798	−0.03488	−0.5031	1.000	0.3369	−0.7299	−0.3495
	2.051	1.000	−0.1349	0.2872	0.1120	0.6444	−0.3216
	8.100	−0.04175	0.1631	0.8390	−0.5291	0.3161	1.000
	10.09	−0.4616	0.7252	0.3558	1.000	0.5028	−0.06588

Continuation of 108

k	λ_i	x_1	x_2	x_3	x_4	x_5	x_6
1	− 7.186	0.4503	−0.4540	−0.3950	0.9123	−0.3191	1.000
	− 2.661	0.6285	1.000	0.03761	−0.07126	−0.7464	0.01276
	− 0.3768	−0.02062	−0.5179	1.000	0.3334	−0.6988	−0.3580
	2.232	1.000	−0.1402	0.2424	0.1262	0.6487	−0.3264
	8.392	−0.01954	0.1913	0.8647	−0.4953	0.3477	1.000
	10.40	−0.4546	0.7087	0.3276	1.000	0.4860	−0.1014
2	− 6.782	0.4175	−0.4733	−0.3979	0.9190	−0.3157	1.000
	− 2.359	0.6547	1.000	0.04177	−0.04554	−0.7738	0.01418
	− 0.07544	−0.0046	−0.5341	1.000	0.3307	−0.6663	−0.3672
	2.410	1.000	−0.1450	0.1965	0.1395	0.6550	−0.3294
	8.689	0.00237	0.2195	0.8895	−0.4608	0.3791	1.000
	10.72	−0.4501	0.6929	0.3004	1.000	0.4694	−0.1353
3	− 6.386	0.3828	−0.4927	−0.4007	0.9242	−0.3114	1.000
	− 2.056	0.6832	1.000	0.04632	−0.01758	−0.8037	0.01573
	0.2241	0.01345	−0.5519	1.000	0.3291	−0.6322	−0.3775
	2.588	1.000	−0.1490	0.1490	0.1520	0.6634	−0.3305
	8.993	0.02384	0.2479	0.9134	−0.4255	0.4103	1.000
	11.04	−0.4479	0.6777	0.2739	1.000	0.4531	−0.1680
4	− 5.998	0.3463	−0.5119	−0.4032	0.9278	−0.3060	1.000
	− 1.752	0.7142	1.000	0.05133	0.01294	−0.8364	0.01740
	0.5213	0.03373	−0.5717	1.000	0.3288	−0.5964	−0.3887
	2.765	1.000	−0.1519	0.09946	0.1636	0.6740	−0.3295
	9.302	0.04461	0.2769	0.9367	−0.3891	0.4414	1.000
	11.36	−0.4478	0.6629	0.2482	1.000	0.4368	−0.1997
5	− 5.620	0.3079	−0.5308	−0.4054	0.9298	−0.2995	1.000
	− 1.449	0.7480	1.000	0.05686	0.04636	−0.8722	0.01921
	0.8158	0.05651	−0.5938	1.000	0.3300	−0.5585	−0.4012
	2.944	1.000	−0.1534	0.04752	0.1743	0.6869	−0.3264
	9.616	0.06453	0.3064	0.9594	−0.3514	0.4725	1.000
	11.69	−0.4496	0.6485	0.2228	1.000	0.4206	−0.2308

109.

k	λ_i	x_1	x_2	x_3	x_4	x_5	x_6	x_7
0	−5.6067	0.61472	0.64995	−0.70581	−0.55160	1.0000	0.057260	−0.40586
	−4.9023	0.18033	0.05290	−0.81318	0.95494	0.23530	−0.38216	1.0000
	−2.6995	1.0000	−0.93848	−0.37965	−0.24449	−0.41072	0.42119	0.051678
	−0.54980	−0.44839	0.19060	−0.45610	0.12950	−0.15782	1.0000	−0.004476
	4.2478	−0.57270	−0.82789	0.21415	−0.60648	1.0000	0.23841	0.75618
	5.3502	−0.06898	−0.67365	0.08140	1.0000	0.69272	0.11012	−0.96159
	8.9903	0.73105	0.46130	1.0000	0.41165	0.32760	0.69639	0.45289
1	−4.7400	0.75585	0.58775	−0.79150	−0.54757	1.0000	0.033716	−0.28069
	−3.1870	−0.16636	0.21407	−0.71934	1.0000	0.25996	−0.43116	0.95223
	−1.8735	−0.97270	1.0000	0.46001	0.086437	0.46016	−0.26481	−0.38355
	0.37011	−0.36858	0.14561	−0.45598	0.12684	−0.15456	1.0000	−0.079792
	5.0743	−0.51769	−0.36572	0.23304	−0.84187	0.60204	0.24833	1.0000
	5.7829	−0.22262	−0.97799	0.11073	0.62825	1.0000	0.13736	−0.60596
	10.403	0.78632	0.45568	1.0000	0.55272	0.33049	0.69509	0.43442
2	−3.9721	0.94960	0.51245	−0.83591	−0.68659	1.0000	−0.043198	−0.22313
	−1.8966	−0.64680	0.74620	−0.60943	1.0000	0.57103	−0.47516	0.63434
	−0.74211	0.78734	−0.97942	−0.79445	0.37194	−0.43170	0.008066	1.0000
	1.2905	−0.28887	0.10093	−0.45401	0.11985	−0.15188	1.0000	−0.15261
	5.6499	−0.42614	0.11468	0.21587	−0.84145	0.16131	0.24128	1.0000
	6.5868	−0.32318	−0.90704	0.17393	0.17504	1.0000	0.18812	−0.13066
	11.914	0.85635	0.45635	1.0000	0.68017	0.32626	0.69196	0.44946
3	−3.3485	1.0000	0.32782	−0.71173	−0.75635	0.80635	0.049705	−0.11874
	−1.0575	−0.70305	1.0000	−0.53927	0.78948	0.78465	−0.36965	0.21723
	0.75068	0.34403	−0.61429	−0.77145	0.48480	−0.24982	−0.09791	1.0000
	2.2160	−0.21255	0.06071	−0.44615	0.10733	−0.14901	1.0000	−0.22511
	6.2018	−0.46606	0.35168	0.24747	−0.82860	0.004692	0.30474	1.0000
	7.5519	−0.35730	−0.85778	0.22116	0.024528	1.0000	0.22722	0.026798
	13.516	0.93830	0.45952	1.0000	0.80776	0.31908	0.68801	0.48639

110. $p_1 = -0.215859$, $p_2 = 1.30441$, $p_3 = 0.141684$, $p_4 = -0.410047$.
111. $p_1 = -0.169728$, $p_2 = 1.31243$, $p_3 = 0.115378$, $p_4 = -0.411901$.
112. $p_1 = -0.187235$, $p_2 = 1.31628$, $p_3 = 0.123431$, $p_4 = -0.414992$.
113. $p_1 = -0.175228$, $p_2 = 1.28112$, $p_3 = 0.109620$, $p_4 = -0.394360$.

114.

k	p_1	p_2	p_3
0	12	63	−308
1	15	53	−385
2	18	39	−470
3	21	21	−563
4	24	− 1	−664
5	27	−27	−773

115.

k	p_1	p_2	p_3
0.0	19	24	1
0.1	19.2	24.11	1.208
0.2	19.4	24.24	1.024
0.3	19.6	24.39	0.436
0.4	19.8	24.56	− 0.568
0.5	20	24.75	− 2
0.6	20.2	24.96	− 3.872
0.7	20.4	25.19	− 6.196
0.8	20.6	25.44	− 8.984
0.9	20.8	25.71	−12.248
1.0	21	26	−16
1.1	21.2	26.31	−20.252
1.2	21.4	26.64	−25.016
1.3	21.6	26.99	−30.304
1.4	21.8	27.36	−36.128
1.5	22	27.75	−42.5

116.

k	p_1	p_2	p_3	p_4
0	18.2	− 4.82	−695.58	1077.6
1	18.6	− 9.57	−698.73	1100.7
2	19	−14.42	−701.25	1123.6
3	19.4	−19.37	−703.14	1146.2
4	19.8	−24.42	−704.39	1168.5
5	20.2	−29.57	−704.98	1190.5

117

k	p_1	p_2	p_3	p_4
0.0	13.7	107.28	−407	−1107.4
0.1	14.1	104.26	−433.48	−1065.9
0.2	14.5	101.14	−459.78	−1020.6
0.3	14.9	97.92	−485.91	− 971.67
0.4	15.3	94.6	−511.84	− 918.94
0.5	15.7	91.18	−537.56	− 862.41
0.6	16.1	87.66	−563.06	− 802.01
0.7	16.5	84.04	−588.32	− 737.71
0.8	16.9	80.32	−613.35	− 669.45
0.9	17.3	76.5	−638.12	− 597.18
1.0	17.7	72.58	−662.62	− 520.86
1.1	18.1	68.56	−686.84	− 440.45
1.2	18.5	64.44	−710.76	− 355.88
1.3	18.9	60.22	−734.39	− 267.12
1.4	19.3	55.9	−757.7	− 174.11
1.5	19.7	51.48	−780.68	− 76.808

118.

k	p_1	p_2	p_3	p_4	p_5
0	24.3	5.4	−1540.9	1084.2	1326.9
1	25.3	6.16	−1576.5	− 264.44	4325.9
2	26.3	10.12	−1676.1	−1501.3	8177
3	27.3	17.28	−1846.3	−2553.3	12815
4	28.3	27.64	2094	−3342	18146
5	29.3	41.2	−2425.9	−3783.2	24046
6	30.3	57.96	−2848.6	−3787.2	30358
7	31.3	77.92	−3368.8	−3258.6	36890
8	32.3	101.08	−3993.4	−2096.8	43414
9	33.3	127.44	−4729	− 195.27	49665
10	34.3	157	−5582.3	2558	55337

119.

k	p_1	p_2	p_3	p_4	p_5	p_6
0	6.25	1.1107	− 62.147	94.720	−− 15.715	6.1555
1	9.25	−11.264	−101.84	310.10	− 228.01	0.33127
2	12.25	−27.139	−161.18	766.68	− 935.68	204.72
3	15.25	−46.514	−249.17	1635.4	− 2778.3	1177.6
4	18.25	−69.389	−374.81	3136.6	− 6800.7	4171.5
5	21.25	−95.764	−547.10	5540.3	−14550	11496

120.

k	p_1	p_2	p_3	p_4	p_5	p_6	p_7
0	25	176	−8598	46094	$31293 \cdot 10$	$-27087 \cdot 10^2$	$31672 \cdot 10^2$
1	26	149	−8660	49525	$29910 \cdot 10$	$-27070 \cdot 10^2$	$29780 \cdot 10^2$
2	28	125	−8736	52361	$29014 \cdot 10$	$-26922 \cdot 10^2$	$26601 \cdot 10^2$
3	29	102	−8837	54681	$28708 \cdot 10$	$-26710 \cdot 10^2$	$22075 \cdot 10^2$

121. $p_1 = 18.737, p_2 = -110.591, p_3 = 212.402.$ **122.** $p_1 = 97, p_2 = -77,$ $p_3 = -457, p_4 = 450.$

123. $\lambda_1 = -0.943568,\ \lambda_2 = -0.744036,\ \lambda_3 = 0.687843,\ \lambda_4 = 0.857774;$

$$X_1 = \begin{bmatrix} 1.00000 \\ -0.458447 \\ -0.056689 \\ -0.834530 \end{bmatrix}, \qquad X_2 = \begin{bmatrix} 0.056689 \\ 0.834530 \\ 1.00000 \\ -0.458447 \end{bmatrix},$$

$$X_3 = \begin{bmatrix} 1.00000 \\ 0.484317 \\ -0.032481 \\ 0.934427 \end{bmatrix}, \qquad X_4 = \begin{bmatrix} 0.032481 \\ -0.934427 \\ 1.00000 \\ 0.484317 \end{bmatrix}.$$

124. $\lambda_1 = -0.947904,\ \lambda_2 = -0.767871,\ \lambda_3 = 0.663188,\ \lambda_4 = 0.856439;$

$$X_1 = \begin{bmatrix} 0.243340 \\ 1.00000 \\ 0.590915 \\ 0.994468 \end{bmatrix}, \qquad X_2 = \begin{bmatrix} -0.590915 \\ -0.994468 \\ 0.243339 \\ 1.00000 \end{bmatrix},$$

$$X_3 = \begin{bmatrix} -0.703487 \\ 0.118487 \\ 1.00000 \\ -0.541209 \end{bmatrix}, \qquad X_4 = \begin{bmatrix} 1.00000 \\ -0.541209 \\ 0.703487 \\ -0.118487 \end{bmatrix}.$$

125. $\lambda_1 = -0.939205,\ \lambda_2 = -0.750904,\ \lambda_3 = 0.680273,\ \lambda_4 = 0.835871;$

$$X_1 = \begin{bmatrix} -0.563913 \\ 1.00000 \\ 0.019513 \\ 0.853597 \end{bmatrix}, \qquad X_2 = \begin{bmatrix} -0.019513 \\ -0.853597 \\ -0.563913 \\ 1.00000 \end{bmatrix},$$

$$X_3 = \begin{bmatrix} 1.00000 \\ 0.550816 \\ -0.808397 \\ 0.033823 \end{bmatrix}, \qquad X_4 = \begin{bmatrix} 0.808397 \\ -0.033822 \\ 1.00000 \\ 0.550816 \end{bmatrix}.$$

126.

k	λ_i	x_1	x_2	x_3
0	−0.19529	−0.20764	−0.42212	1.0000
	0.35958	1.0000	−0.71972	−0.096171
	1.5857	0.77580	1.0000	0.58320
1	−0.55596	−0.98814	1.0000	−0.23778
	0.046539	−0.32851	−0.086835	1.0000
	2.5594	0.91850	1.0000	0.38857
2	−1.4447	1.0000	−0.98000	0.095657
	0.19216	−0.19181	−0.098119	1.0000
	3.6025	0.95313	1.0000	0.28094

Continuation of 126

k	λ_i	x_1	x_2	x_3
3	−2.3415	1.0000	−0.98110	0.061350
	0.32066	−0.14131	−0.081501	1.0000
	4.6708	0.96771	1.0000	0.21825
4	−3.2399	1.0000	−0.98363	0.045436
	0.43907	−0.11250	−0.068179	1.0000
	5.7508	0.97555	1.0000	0.17793
5	−4.1390	1.0000	−0.98582	0.036154
	0.55182	−0.093580	−0.058252	1.0000
	6.8371	0.98039	1.0000	0.15000
6	−5.0383	1.0000	−0.98756	0.030049
	0.66114	−0.080146	−0.050728	1.0000
	7.9272	0.98366	1.0000	0.12956
7	−5.9379	1.0000	−0.98894	0.025719
	0.76823	−0.070098	−0.044874	1.0000
	9.0196	0.98601	1.0000	0.11399

127.

k	λ_i	x_1	x_2	x_3
0	−0.90864	1.0000	−0.87716	0.96619
	1.2527	−0.013250	1.0000	0.92158
	2.7559	1.0000	0.52654	−0.55697
1	−0.77032	0.92787	−0.91612	1.0000
	1.5357	−0.16400	0.92546	1.0000
	2.7346	1.0000	0.59290	−0.38470
2	−0.65436	0.81000	0.90669	1.0000
	1.7524	−0.37825	0.76500	1.0000
	2.8019	1.0000	0.71081	−0.16552
3	−0.56413	0.68651	−0.88357	1.0000
	1.8835	−0.62745	0.64426	1.0000
	2.9807	1.0000	0.86001	0.073379
4	−0.50304	0.56292	−0.85132	1.0000
	1.9530	−0.87421	0.59659	1.0000
	3.2500	1.0000	0.99255	0.28206
5	−0.47369	0.44453	−0.81452	1.0000
	2.0018	1.0000	−0.55623	−0.89759
	3.5719	0.92018	1.0000	0.40548

Continuation of 127

k	λ_i	x_1	x_2	x_3
6	−0.47699	0.33558	−0.77715	1.0000
	2.0557	1.0000	−0.51491	−0.73574
	3.9213	0.87151	1.0000	0.48468
7	−0.51178	0.23867	−0.74202	1.0000
	2.1259	1.0000	−0.51029	−0.61732
	4.2859	0.84400	1.0000	0.54059

128.

k	λ_i	x_1	x_2	x_3
0	−1.8122	1.0000	0.96013	0.90681
	2.1650	−0.42878	−0.49788	1.0000
	2.4953	1.0000	−0.98385	−0.060065
1	−0.81551	1.0000	0.95584	0.86240
	1.6685	−0.43168	−0.45062	1.0000
	2.9951	−0.97972	1.0000	0.027696
2	0.17177	1.0000	0.95288	0.70376
	1.1811	−0.35324	−0.36785	1.0000
	3.4953	−0.97050	1.0000	0.025036
3	0.62407	−0.82828	−0.82341	1.0000
	1.2285	0.63032	0.58041	1.0000
	3.9956	−0.96244	1.0000	0.026240
4	0.15020	−0.57725	−0.57961	1.0000
	2.2019	0.90070	0.82827	1.0000
	4.4960	0.95260	1.0000	0.029732
5	−0.34553	−0.54300	−0.54561	1.0000
	3.1970	0.96654	0.87089	1.0000
	4.9966	−0.93837	1.0000	0.036080
6	−0.84380	−0.52964	−0.53208	1.0000
	4.1941	1.0000	0.86704	0.99098
	5.4978	−0.91439	1.0000	0.047777
7	−1.3429	−0.52257	−0.52478	1.0000
	5.1906	1.0000	0.79438	0.93944
	6.0004	−0.86348	1.0000	0.073550

129.

k	λ_i	x_1	x_2	x_3	x_4
0	−49.240	1.0000	−0.43469	−0.26135	−0.23760
	−18.490	−0.003109	−0.16895	−0.64002	1.0000
	38.100	0.45112	0.078677	1.0000	0.65471
	67.631	0.37098	1.0000	−0.25186	0.008903
1	−48.621	1.0000	−0.43118	−0.25130	−0.26272
	−17.480	0.017920	−0.18519	−0.65641	1.0000
	39.462	0.45461	0.068779	1.0000	0.66100
	68.638	0.37280	1.0000	−0.24845	0.015420
2	−48.037	1.0000	−0.42757	−0.24143	−0.28710
	−16.442	0.038508	−0.20134	−0.67309	1.0000
	40.829	0.45802	0.058534	1.0000	0.66724
	69.650	0.37478	1.0000	−0.24491	0.022064
3	−47.488	1.0000	−0.42389	−0.23177	−0.31069
	−15.378	0.058605	−0.21740	−0.69005	1.0000
	42.199	0.46132	0.047929	1.0000	0.67343
	70.666	0.37694	1.0000	−0.24124	0.028841
4	−46.971	1.0000	−0.42015	−0.22237	−0.33348
	−14.289	0.078166	−0.23333	−0.70725	1.0000
	43.573	0.46452	0.036953	1.0000	0.67956
	71.686	0.37928	1.0000	−0.23744	0.035761
5	−46.485	1.0000	−0.41637	−0.21325	−0.35543
	−13.178	0.097160	−0.24913	−0.72467	1.0000
	44.951	0.46759	0.025593	1.0000	0.68562
	72.711	0.38180	1.0000	−0.23349	0.042835

130.

k	λ_i	x_1	x_2	x_3	x_4
0	−17.137	0.81812	1.0000	−0.99008	−0.98398
	4.6240	1.0000	−0.41590	0.23739	0.16990
	21.929	0.16842	1.0000	0.76797	0.38358
	34.584	0.081276	0.17706	−0.74785	1.0000
1	−17.557	−0.88739	−0.98310	1.0000	0.99521
	5.7566	1.0000	−0.48869	0.23586	0.17191
	22.689	0.23265	1.0000	0.80095	0.39048
	35.112	0.090155	0.17991	−0.73833	1.0000
2	−18.026	−0.94037	−0.95525	1.0000	0.99621
	6.8528	1.0000	−0.56348	0.23119	0.17156
	23.529	0.30067	1.0000	0.84120	0.39831
	35.644	0.099866	0.18335	−0.72715	1.0000

Continuation of 130

k	λ_i	x_1	x_2	x_3	x_4
3	−18.537	−0.98640	−0.92708	1.0000	0.99691
	7.9048	1.0000	−0.64056	0.22381	0.16927
	24.450	0.37268	1.0000	0.88948	0.40646
	36.181	0.11061	0.18744	−0.71404	1.0000
4	−19.083	1.0000	0.87582	−0.97419	−0.97165
	8.9075	1.0000	−0.72002	0.21415	0.16546
	25.450	0.44877	1.0000	0.94669	0.41407
	36.726	0.12262	0.19230	−0.69864	1.0000
5	−19.662	1.0000	0.82091	−0.94204	−0.93987
	9.8584	1.0000	−0.80180	0.20267	0.16053
	26.525	0.52163	0.98626	1.0000	0.41412
	37.278	0.13620	0.19806	−0.68052	1.0000
6	−20.267	1.0000	0.77316	−0.91556	−0.91360
	10.758	1.0000	−0.88569	0.18980	0.15482
	27.669	0.56098	0.91522	1.0000	0.38643
	37.841	0.15175	0.20490	−0.65909	1.0000
7	−20.898	1.0000	0.73116	−0.89346	−0.89161
	11.607	1.0000	−0.97142	0.17593	0.14866
	28.874	0.59144	0.84403	1.0000	0.35340
	38.417	0.16976	0.21303	−0.63360	1.0000
8	−21.549	1.0000	0.69385	−0.87482	−0.87301
	12.409	−0.94460	1.0000	−0.15250	−0.13441
	30.129	0.61245	0.77322	1.0000	0.31400
	39.011	0.19086	0.22272	0.60311	1.0000
9	−22.220	1.0000	0.66043	−0.85895	−0.85712
	13.169	−0.87180	1.0000	−0.12783	−0.11850
	31.423	0.62351	0.70304	1.0000	0.26702
	39.629	0.21589	0.23432	−0.56637	1.0000
10	−22.909	1.0000	0.63028	−0.84531	−0.84342
	13.890	−0.80886	1.0000	−0.10655	−0.10493
	32.741	0.62411	0.63353	1.0000	0.21116
	40.278	0.24587	0.24820	−0.52185	1.0000
11	−23.613	1.0000	0.60291	−0.83351	−0.83154
	14.578	−0.75405	1.0000	−0.088192	−0.093360
	34.064	0.61381	0.56459	1.0000	0.14515
	40.970	0.28203	0.26477	−0.46775	1.0000

Continuation of 130

k	λ_i	x_1	x_2	x_3	x_4
12	−24.330	1.0000	0.57793	−0.82322	−0.82115
	15.237	−0.70603	1.0000	−0.072324	−0.083483
	35.371	0.59241	0.49627	1.0000	0.068193
	41.723	0.32569	0.28434	−0.40224	1.0000
13	−25.061	1.0000	0.55503	−0.81419	−0.81202
	15.871	−0.66367	1.0000	−0.058587	−0.075047
	36.635	0.56036	0.42903	1.0000	−0.019350
	42.555	0.37788	0.30685	−0.32404	1.0000
14	−25.803	1.0000	0.53393	−0.80623	−0.80395
	16.483	−0.62609	1.0000	−0.046669	−0.067831
	37.828	0.51937	0.36403	1.0000	−0.11505
	43.492	0.43867	0.33158	−0.23348	1.0000
15	−26.556	1.0000	0.51443	−0.79917	−0.79677
	17.077	−0.59256	1.0000	−0.036303	−0.061647
	38.927	0.47268	0.30320	1.0000	−0.21401
	44.552	0.50643	0.35689	−0.13357	1.0000

131.

k	λ_i	x_1	x_2	x_3	x_4	x_5
0	−0.78913	−0.48631	1.0000	−0.88383	−0.57692	0.87338
	−0.39233	1.0000	−0.30945	−0.84568	0.085722	0.11194
	0.046419	−0.32650	−0.72168	0.056194	0.45207	1.0000
	0.36766	−0.46586	0.20616	−0.58090	1.0000	−0.42274
	2.7674	0.78723	1.0000	0.72320	0.82045	0.56717
1	−0.78872	1.0000	−0.63782	−0.58959	0.15271	−0.025453
	0.25351	−0.000408	−0.77120	1.0000	0.51702	−0.75274
	1.1527	−0.12599	−0.55082	0.33530	−0.014310	1.0000
	1.5805	−0.34467	0.048996	−0.38405	1.0000	0.12664
	4.8020	1.0000	0.96724	0.77262	0.54240	0.40746
2	−1.1816	1.0000	−0.64585	−0.64172	0.10060	0.002841
	1.2559	0.027263	−0.86509	1.0000	0.57593	−0.77462
	2.1741	−0.070428	−0.52355	0.40063	−0.13371	1.0000
	2.7165	−0.25599	0.063191	−0.30454	1.0000	0.27077
	7.0351	1.0000	0.87048	0.73847	0.35081	0.27721
3	−1.5853	1.0000	−0.65707	−0.65984	0.076511	0.007343
	2.2583	0.027056	−0.90174	1.0000	0.60118	−0.77926
	3.1822	−0.048342	−0.51336	0.42782	−0.18325	1.0000
	3.7938	−0.19954	0.082439	−0.26488	1.0000	0.32921
	9.3509	1.0000	0.82511	0.72578	0.25521	0.20818

Continuation of 131

k	λ_i	x_1	x_2	x_3	x_4	x_5
4	−1.9930	1.0000	−0.66526	−0.66972	0.061952	0.008097
	3.2599	0.024324	−0.92229	1.0000	0.61587	−0.78092
	4.1865	−0.036703	−0.50808	0.44256	−0.21001	1.0000
	4.8417	−0.16244	0.096599	−0.24165	1.0000	0.36007
	11.705	1.0000	0.79921	0.71972	0.19934	0.16611
5	−2.4028	1.0000	−0.67126	−0.67608	0.052109	0.007914
	4.2610	0.021631	−0.93560	1.0000	0.62559	−0.78166
	5.1891	−0.029552	−0.50485	0.45178	−0.22670	1.0000
	5.8738	−0.13662	0.10676	−0.22657	1.0000	0.37892
	14.079	1.0000	0.78255	0.71634	0.16310	0.13797
6	−2.8138	1.0000	−0.67579	−0.68054	0.044986	0.007474
	5.2618	0.019331	−0.94497	1.0000	0.63252	−0.78203
	6.1908	−0.024723	−0.50268	0.45809	−0.23810	1.0000
	6.8967	−0.11775	0.11424	−0.21605	1.0000	0.39159
	16.464	1.0000	0.77096	0.71424	0.13782	0.11790
7	−3.2255	1.0000	−0.67932	−0.68387	0.039586	0.006983
	6.2625	0.017413	−0.95195	1.0000	0.63771	−0.78223
	7.1920	−0.021246	−0.50112	0.46267	−0.24638	1.0000
	7.9137	−0.10338	0.11993	−0.20833	1.0000	0.40067
	18.857	1.0000	0.76245	0.71285	0.11923	0.10289
8	−3.6378	1.0000	−0.68214	−0.68644	0.035348	0.006510
	7.2630	0.015815	−0.95734	1.0000	0.64176	−0.78234
	8.1930	−0.018624	−0.49995	0.46615	−0.25265	1.0000
	8.9269	−0.092113	0.12438	−0.20243	1.0000	0.40749
	21.255	1.0000	0.75593	0.71186	0.10501	0.091245
9	−4.0505	1.0000	−0.68444	−0.68850	0.031932	0.006074
	8.2634	0.014470	−0.96164	1.0000	0.64500	−0.78240
	9.1937	−0.016577	−0.49903	0.46889	−0.25758	1.0000
	9.9374	−0.083039	0.12795	−0.19778	1.0000	0.41279
	23.656	1.0000	0.75079	0.71114	0.093797	0.081957
10	−4.4634	1.0000	−0.68636	−0.69018	0.029119	0.005680
	9.2638	0.013328	−0.96515	1.0000	0.64765	−0.78243
	10.194	−0.014934	−0.49829	0.47109	−0.26154	1.0000
	10.946	−0.075581	0.13087	−0.19403	1.0000	0.41703
	26.059	1.0000	0.74662	0.71059	0.084730	0.074377

132.

k	λ_i	x_1	x_2	x_3	x_4	x_5	x_6
0	−7.2445	−0.92174	1.0000	0.96694	−0.63621	−0.054862	−0.56138
	−3.5435	−0.20984	−0.39553	−0.42850	−0.44295	1.0000	0.69384
	1.1550	0.41530	−0.67270	0.90411	0.61571	−0.21260	1.0000
	7.1110	−0.87982	0.095706	−0.36616	1.0000	0.033676	0.15221
	10.892	0.76768	1.0000	−0.15027	0.46152	0.48281	0.30820
	14.130	−0.099449	−0.17539	0.84743	0.33305	1.0000	−0.83533
1	−6.2498	−0.94266	0.98493	1.0000	−0.57742	−0.034271	0.67464
	−2.5997	−0.20831	−0.42805	−0.44645	−0.39922	1.0000	0.70473
	2.5756	0.33881	−0.83294	0.90520	0.77849	−0.27578	1.0000
	7.1125	−0.92356	0.13054	−0.46737	1.0000	0.006182	0.067922
	12.472	0.82872	1.0000	−0.065214	0.57908	0.57749	0.31965
	15.189	−0.20012	−0.24795	0.87535	0.31146	1.0000	−0.89777
2	−5.3463	−0.96247	0.94193	1.0000	−0.50199	−0.004675	0.77641
	−1.6652	−0.20604	−0.46822	−0.47653	−0.34987	1.0000	0.70620
	3.8597	0.17875	−0.96605	0.83672	1.0000	−0.34512	0.96038
	7.2526	−0.99323	0.24082	−0.63173	1.0000	0.001570	−0.063186
	14.087	0.90600	1.0000	0.075357	0.72350	0.74258	0.28513
	16.312	0.35481	0.36779	−0.90938	−0.24592	−0.98027	1.0000
3	−4.5409	1.0000	−0.89186	−0.98759	0.41974	−0.030910	−0.87902
	−0.73813	−0.20228	−0.51544	−0.51798	−0.29592	1.0000	0.69833
	4.9191	−0.062659	−0.79312	0.50332	1.0000	−0.32351	0.65681
	7.6313	1.0000	−0.44106	0.81875	−0.87377	−0.033544	0.24919
	15.665	0.93795	0.94995	0.30815	0.88557	1.0000	0.14470
	17.563	0.52837	0.48862	−0.80710	−0.083002	−0.78223	1.0000

Continuation of 132

k	λ_i	x_1	x_2	x_3	x_4	x_5	x_6
4	−3.8441	1.0000	−0.79233	−0.91400	0.31077	−0.065979	−0.93103
	0.18436	−0.19642	−0.56855	−0.56956	−0.23901	1.0000	0.68137
	5.6837	−0.22991	−0.62192	0.23070	1.0000	−0.30875	0.41152
	8.3350	0.95875	−0.70960	1.0000	−0.66336	−0.10241	0.43778
	17.102	0.66070	0.62385	0.46835	0.77847	1.0000	−0.092063
	19.039	0.72470	0.61081	−0.67438	0.12176	−0.54675	1.0000
5	−3.2676	1.0000	−0.68899	−0.82753	0.19955	−0.095823	−0.96300
	1.1056	−0.18809	−0.62575	−0.62920	−0.18130	1.0000	0.65600
	6.2342	−0.30618	−0.50850	0.055813	1.0000	−0.31445	0.23642
	9.3012	0.78898	−0.85140	1.0000	−0.41468	−0.15760	0.49887
	18.358	0.46663	0.41985	0.56443	0.69969	1.0000	−0.25537
	20.769	0.88780	0.68375	−0.54645	0.30356	−0.34995	1.0000

133.

k	λ_i	x_1	x_2	x_3	x_4	x_5	x_6	x_7
0	−52.179	0.043552	−0.51059	0.85904	1.0000	−0.55011	−0.69700	−0.036368
	−36.802	−0.94381	−0.94495	−0.46254	0.027874	0.61154	−0.43166	1.0000
	−0.61985	−0.77138	0.050488	−0.005752	0.24239	−0.96527	1.0000	0.33223
	11.704	0.59489	0.39917	−0.32203	−0.23615	−0.68681	−0.50107	1.0000
	38.187	−0.47010	1.0000	−0.57900	0.90974	0.24983	−0.36222	−0.10104
	69.201	−0.50168	0.65321	1.0000	−0.43972	0.28761	−0.15487	0.37582
	97.809	0.86491	0.043742	0.36109	0.88584	0.87579	1.0000	0.89605
1	−50.966	0.033946	−0.50638	0.85197	1.0000	−0.56293	−0.68741	−0.046788
	−35.799	−0.94374	−0.94830	−0.45784	0.031566	0.60820	−0.43568	1.0000
	0.11785	−0.77565	0.059291	−0.023091	0.25155	−0.93512	1.0000	0.31012
	12.588	0.58276	0.38764	−0.30631	−0.23921	−0.69809	−0.48079	1.0000
	38.549	−0.45230	1.000	−0.61558	0.91264	0.24700	−0.38914	−0.10897
	69.735	−0.51700	0.67247	1.0000	−0.41918	0.29185	−0.15587	0.37545
	100.08	0.87182	0.048735	0.36064	0.91620	0.88602	1.0000	0.90199
2	−49.781	0.024125	−0.50166	0.84420	1.0000	−0.57546	−0.67713	−0.057319
	−34.795	−0.94385	−0.95178	−0.45314	0.035398	0.60472	−0.43973	1.0000
	0.83723	−0.77997	0.068341	−0.041029	0.26060	−0.90498	1.0000	0.28804
	13.470	0.57021	0.37636	−0.29074	−0.24192	−0.70948	−0.45996	1.0000
	38.903	−0.43312	1.0000	−0.65162	0.91430	0.24458	−0.41649	−0.11571
	70.301	−0.53274	0.69165	1.0000	−0.39978	0.29558	−0.15797	0.37455
	102.36	0.87891	0.053884	0.36163	0.94579	0.89675	1.0000	0.90868
3	−48.625	0.014080	−0.49642	0.83571	1.0000	−0.58766	−0.66617	−0.067936
	−33.790	−0.94416	−0.95535	−0.44845	0.039359	0.60111	−0.44377	1.0000
	1.5377	−0.78428	0.077665	−0.059592	0.26951	−0.87488	1.0000	0.26605
	14.351	0.55728	0.36536	−0.27540	−0.24426	−0.72092	−0.43867	1.0000
	39.253	−0.41261	1.0000	−0.68694	0.91466	0.24254	−0.44426	−0.12122
	70.898	−0.54872	0.71066	1.0000	−0.38144	0.29887	−0.16097	0.37322
	104.68	0.88620	0.059120	0.36383	0.97475	0.90789	1.0000	0.91602

134. $\lambda_1 = -0.955936$,

$$X = \begin{bmatrix} 1.00000 \\ 0.496884 \\ -0.307464 \\ -0.575794 \end{bmatrix}.$$

135. $\lambda_1 = -0.939092$,

$$X = \begin{bmatrix} -0.824755 \\ 0.199269 \\ 0.453853 \\ 1.00000 \end{bmatrix}.$$

136. $\lambda_1 = -0.953090$,

$$X = \begin{bmatrix} 0.660274 \\ 1.00000 \\ 0.365107 \\ 0.546294 \end{bmatrix}.$$

137. $\lambda_1 = -0.938896$,

$$X = \begin{bmatrix} 1.00000 \\ -0.456253 \\ 0.142602 \\ -0.634066 \end{bmatrix}.$$

138.

k	λ_1	x_1	x_2	x_3
0	14.846	0.75409	1.0000	0.18661
1	16.791	0.77477	1.0000	0.18199
2	18.743	0.79197	1.0000	0.17678
3	20.701	0.80654	1.0000	0.17133
4	22.663	0.81908	1.0000	0.16586
5	24.629	0.82999	1.0000	0.16047

139.

k	λ_1	x_1	x_2	x_3
0.0	20.191	1.0000	0.80348	0.68046
0.1	20.386	1.0000	0.80404	0.68356
0.2	20.580	1.0000	0.80460	0.68659
0.3	20.775	1.0000	0.80516	0.68956
0.4	20.970	1.0000	0.80572	0.69245
0.5	21.165	1.0000	0.80628	0.69529
0.6	21.360	1.0000	0.80685	0.69806
0.7	21.555	1.0000	0.80742	0.70077
0.8	21.751	1.0000	0.80799	0.70343
0.9	21.946	1.0000	0.80855	0.70603
1.0	22.142	1.0000	0.80913	0.70858
1.1	22.337	1.0000	0.80970	0.71107
1.2	22.533	1.0000	0.81027	0.71352
1.3	22.729	1.0000	0.81084	0.71592
1.4	22.925	1.0000	0.81141	0.71827
1.5	23.121	1.0000	0.81198	0.72057

140.

k	λ_1	x_1	x_2	x_3	x_4
0	15.167	0.82007	0.56902	0.55665	1.0000
1	15.295	0.82224	0.57624	0.56412	1.0000
2	15.423	0.82445	0.58360	0.57173	1.0000
3	15.552	0.82669	0.59110	0.57947	1.0000
4	15.681	0.82897	0.59874	0.58736	1.0000
5	15.811	0.83129	0.60654	0.59540	1.0000

141.

k	λ_1	x_1	x_2	x_3	x_4
0.0	18.185	1.0000	0.59340	0.92388	0.49879
0.1	18.325	1.0000	0.59245	0.92099	0.50276
0.2	18.467	1.0000	0.59149	0.91809	0.50669
0.3	18.608	1.0000	0.59051	0.91517	0.51059
0.4	18.750	1.0000	0.58951	0.91224	0.51446
0.5	18.892	1.0000	0.58850	0.90930	0.51830
0.6	19.035	1.0000	0.58746	0.90635	0.52211
0.7	19.178	1.0000	0.58641	0.90338	0.52588
0.8	19.321	1.0000	0.58535	0.90040	0.52962
0.9	19.465	1.0000	0.58427	0.89741	0.53333
1.0	19.609	1.0000	0.58317	0.89442	0.53701
1.1	19.753	1.0000	0.58206	0.89141	0.54066
1.2	19.898	1.0000	0.58093	0.88839	0.54428
1.3	20.043	1.0000	0.57979	0.88537	0.54787
1.4	20.189	1.0000	0.57864	0.88234	0.55142
1.5	20.335	1.0000	0.57747	0.87930	0.55495

142.

k	λ_1	x_1	x_2	x_3	x_4	x_5
0	21.266	1.0000	0.95610	0.56427	0.44901	0.87160
1	22.437	1.0000	0.94536	0.55636	0.42389	0.80954
2	23.642	1.0000	0.93582	0.55159	0.40093	0.75524
3	24.879	1.0000	0.92714	0.54908	0.37986	0.70730
4	26.146	1.0000	0.91911	0.54821	0.36049	0.66465
5	27.439	1.0000	0.91161	0.54854	0.34264	0.62646
6	28.756	1.0000	0.90456	0.54973	0.32618	0.59209
7	30.095	1.0000	0.89790	0.55154	0.31097	0.56100
8	31.454	1.0000	0.89160	0.55380	0.29690	0.53277
9	32.832	1.0000	0.88563	0.55638	0.28386	0.50704
10	34.226	1.0000	0.87996	0.55916	0.27177	0.48350

143.

k	λ_1	x_1	x_2	x_3	x_4	x_5	x_6
0	4.2575	0.31592	0.71726	1.0000	0.29645	0.76597	0.90309
1	5.2644	0.92419	0.50948	0.80880	0.73807	0.89017	1.0000
2	6.7433	1.0000	0.21944	0.40995	0.77709	0.58827	0.79710
3	8.4441	1.0000	0.11759	0.25668	0.76209	0.42216	0.73058
4	10.236	1.0000	0.074399	0.18558	0.75177	0.32836	0.70966
5	12.072	1.0000	0.052039	0.14519	0.74455	0.26823	0.70183

144.

k	λ_1	x_1	x_2	x_3	x_4	x_5	x_6	x_7
0	−17.328	0.17426	−0.96247	1.0000	−0.92631	0.93614	0.17595	−0.56428
1	17.719	0.81339	−0.27148	0.36333	1.0000	0.19883	0.11982	0.14895
2	18.954	0.85704	−0.27244	0.37970	1.0000	0.18629	0.12166	0.24724
3	20.249	0.87734	−0.27001	0.37924	1.0000	0.18027	0.12098	0.31344

145. $\lambda_1 = 9.3052,$

$$X = \begin{bmatrix} 1.00000 \\ 0.79880 \\ 0.42394 \end{bmatrix}.$$

146. $\lambda_1 = 96.150,$

$$X = \begin{bmatrix} 0.094743 \\ 0.25986 \\ 1.0000 \\ 0.55005 \end{bmatrix}.$$

ANSWERS TO THE PROBLEMS OF CHAPTER 3

1. a–d)

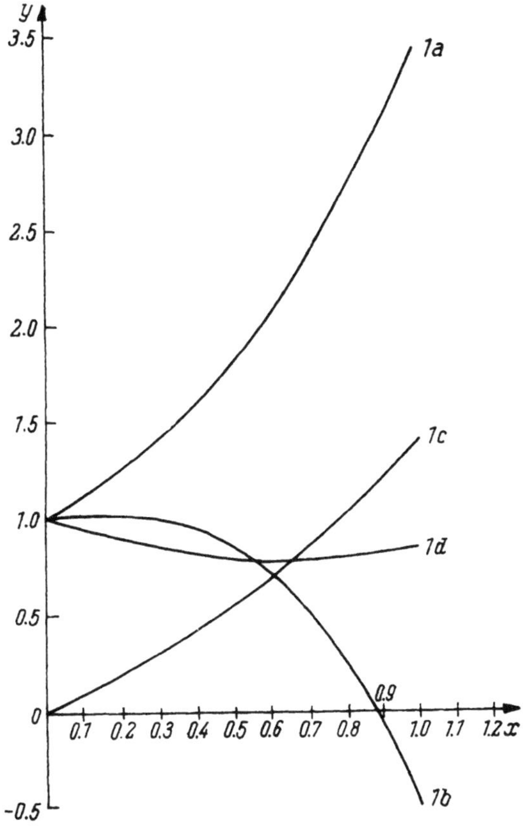

Fig. 17

1. e)

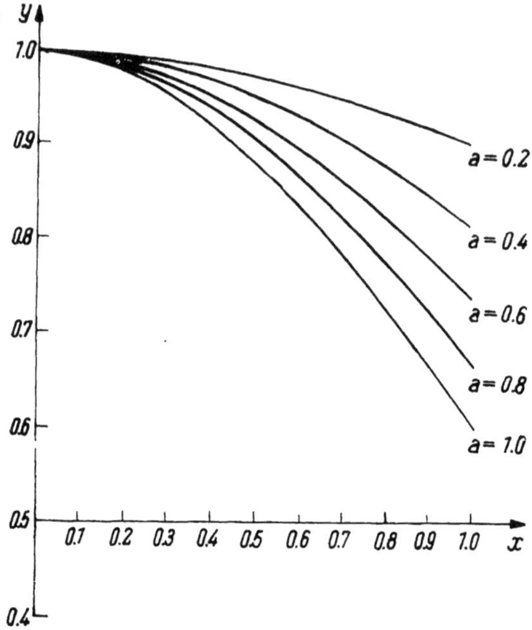

Fig. 18

1. f)

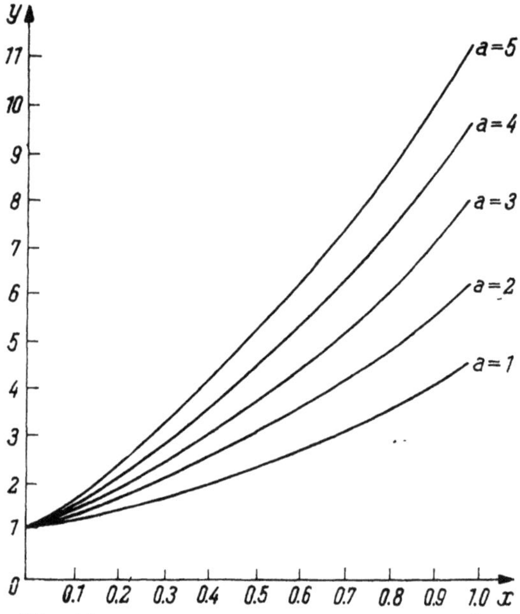

Fig. 19

1. g)

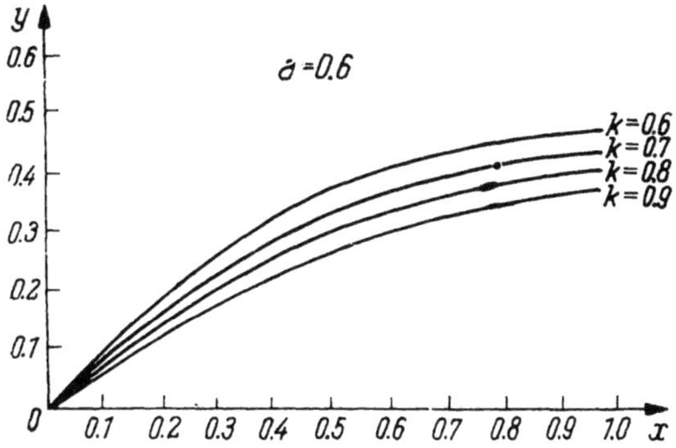

Fig. 20

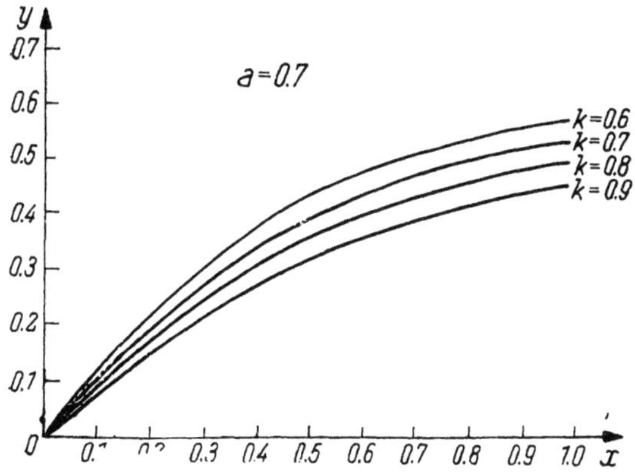

Fig. 21

1. g)

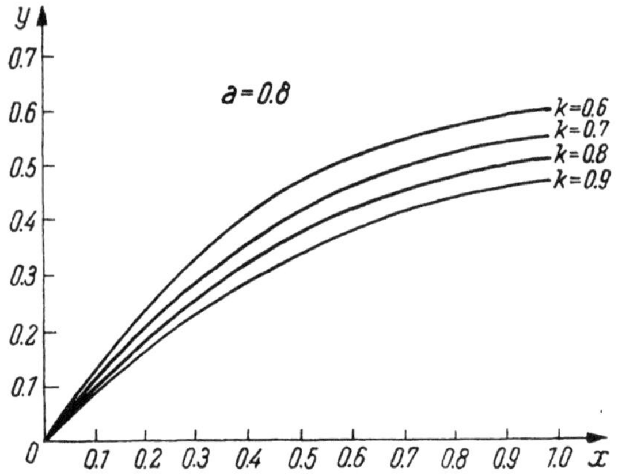

Fig. 22

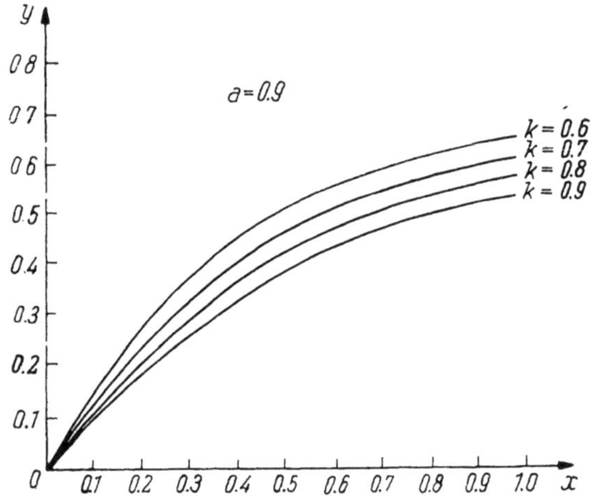

Fig. 23

2. $y(x) = 1 + x + x^2 + \frac{x^3}{3} + \frac{x^4}{3 \cdot 4} + \frac{x^5}{3 \cdot 4 \cdot 5} + \frac{x^6}{3 \cdot 4 \cdot 5 \cdot 6} + \cdots$

x	−0.2	−0.1	0.1	0.2
y	0.83746	0.90967	1.11034	1.24281

3. a) $y(x) = \frac{x^2}{2!} - \frac{x^3}{3!} + \frac{x^4}{4!} - \frac{x^5}{5!} + \frac{x^6}{6!} - \cdots;$

b) $y(x) = 1 + x - \frac{x^2}{2!} \cdot 1 + \frac{x^3}{3!} 1 \cdot 3 - \frac{x^4}{4!} 1 \cdot 3 \cdot 5 + \frac{x^5}{5!} 1 \cdot 3 \cdot 5 \cdot 7 - \cdots;$

c) $y(x) = 1 + xa + \frac{x^2}{2!} a(a-1) + \frac{x^3}{3!} a(a-1)(a-2) +$

$+ \frac{x^4}{4!} a(a-1)(a-2)a-3) + \frac{x^5}{5!} a(a-1)(a-2)(a-3)(a-4) + \cdots;$

	a	x			
		−0.2	−0.1	0.1	0.2
a)		0.02140	0.00517	0.00484	0.01873
b)		0.77460	0.89443	1.09545	1.18322
c)	1	0.80000	0.90000	1.10000	1.20000
	2	0.64000	0.81000	1.21000	1.44000
	3	0.51200	0.72900	1.33100	1.72800
	4	0.40960	0.65610	1.46410	2.07360
	5	0.32768	0.59049	1.61051	2.48832

4.

	x			
	0.05	0.10	0.15	0.20
a)	2.1013	2.2052	2.3118	2.4213
b)	0.9975	0.9900	0.9775	0.9600

5.

	a	x			
		−0.025	−0.050	−0.075	−0.100
a)		−0.02501	−0.05004	−0.07514	−0.10033
b)	0.2	0.99994	0.99975	0.99944	0.99900
	0.4	0.99988	0.99950	0.99888	0.99800
	0.6	0.99981	0.99925	0.99831	0.99700
	0.8	0.99975	0.99900	0.99775	0.99600
	1.0	0.99969	0.99875	0.99719	0.99500

6.

	x			
	0.1	0.2	0.3	0.4
a)	0.099668	0.19738	0.29131	0.37995
b)	−0.95206	−0.90654	−0.86113	−0.81371

7.

	a	x			
		−0.05	−0.10	−0.15	−0.20
a)		0.0012083	0.0046662	0.010122	0.017321
b)	1	0.90246	0.80968	0.72142	0.63746
	2	0.85369	0.71451	0.58212	0.45619
	3	0.80492	0.61935	0.44283	0.27492
	4	0.75615	0.52419	0.30354	0.093653
	5	0.70738	0.42902	0.16425	−0.087616

8. a)

a	x	k				
		1.00	1.25	1.50	1.75	2.00
0.2	0.2	0.1992	0.1989	0.1987	0.1984	0.1982
	0.4	0.3890	0.3857	0.3824	0.3791	0.3759
	0.6	0.5587	0.5470	0.5357	0.5248	0.5143
	0.8	0.7018	0.6763	0.6526	0.6306	0.6102
	1.0	0.8164	0.7738	0.7359	0.7020	0.6715
0.4	0.4	0.3917	0.3883	0.3850	0.3817	0.3785
	0.7	0.6490	0.6303	0.6126	0.5959	0.5801
	1.0	0.8541	0.8084	0.7677	0.7314	0.6989
0.6	0.5	0.4881	0.4810	0.4742	0.4675	0.4610
	1.0	0.8934	0.8444	0.8009	0.7621	0.7273
0.8	1.0	0.9346	0.8820	0.8355	0.7940	0.7569
1.0	1.0	0.9776	0.9213	0.8715	0.8272	0.7876

8. b)

a	x	k				
		0.50	0.75	1.00	1.25	1.50
1.00	0.2	0.1980	0.1980	0.1980	0.1980	0.1980
	0.4	0.3739	0.3741	0.3744	0.3747	0.3750
	0.6	0.5078	0.5099	0.5119	0.5139	0.5158
	0.8	0.5951	0.6020	0.6086	0.6149	0.6209
	1.0	0.6415	0.6573	0.6722	0.6861	0.6991
1.25	0.4	0.3672	0.3676	0.3679	0.3683	0.3686
	0.7	0.5245	0.5295	0.5343	0.5389	0.5434
	1.0	0.5697	0.5891	0.6071	0.6241	0.6399
1.50	0.5	0.4193	0.4206	0.4220	0.4233	0.4246
	1.0	0.4968	0.5199	0.5414	0.5615	0.5803
1.75	1.0	0.4240	0.4509	0.4761	0.4996	0.5216
2.00	1.0	0.3524	0.3834	0.4124	0.4395	0.4648

8. c)

a	x	k				
		−0.3	−0.1	0.1	0.3	0.5
1.00	0.2	0.1787	0.1797	0.1806	0.1816	0.1825
	0.4	0.2848	0.2892	0.2936	0.2980	0.3025
	0.6	0.3412	0.3497	0.3583	0.3672	0.3764
	0.8	0.3654	0.3781	0.3912	0.4048	0.4190
	1.0	0.3703	0.3871	0.4047	0.4233	0.4429
1.25	0.4	0.2724	0.2767	0.2810	0.2854	0.2898
	0.7	0.3246	0.3349	0.3454	0.3563	0.3676
	1.0	0.3246	0.3409	0.3580	0.3761	0.3952
1.50	0.5	0.2813	0.2874	0.2936	0.2999	0.3064
	1.0	0.2894	0.3053	0.3221	0.3399	0.3587
1.75	1.0	0.2660	0.2817	0.2983	0.3159	0.3345
2.00	1.0	0.2556	0.2712	0.2877	0.3052	0.3238

8. d)

a	x	k				
		−0.5	−0.3	−0.1	0.1	0.3
1.00	0.2	0.1900	0.1902	0.1904	0.1906	0.1908
	0.4	0.3405	0.3428	0.3452	0.3476	0.3500
	0.6	0.4553	0.4628	0.4706	0.4786	0.4869
	0.8	0.5401	0.5557	0.5723	0.5899	0.6085
	1.0	0.6009	0.6267	0.6549	0.6858	0.7199
1.25	0.4	0.2810	0.2826	0.2841	0.2857	0.2873
	0.7	0.4264	0.4343	0.4425	0.4510	0.4599
	1.0	0.5254	0.5445	0.5651	0.5873	0.6114
1.50	0.5	0.2874	0.2897	0.2919	0.2942	0.2965
	1.0	0.4661	0.4808	0.4964	0.5132	0.5311
1.75	1.0	0.4185	0.4301	0.4424	0.4555	0.4693
2.00	1.0	0.3796	0.3890	0.3989	0.4093	0.4203

8. e)

a	x		a	x	
	0.6	1.0		0.6	1.0
14	−0.1215	0.4080	26	−0.0789	0.5004
15	−0.1179	0.4159	27	−0.0753	0.5080
16	−0.1144	0.4237	28	−0.0718	0.5155
17	−0.1108	0.4314	29	−0.0682	0.5230
18	−0.1073	0.4392	30	−0.0647	0.5305
19	−0.1037	0.4469	31	−0.0612	0.5379
20	−0.1001	0.4547	32	−0.0576	0.5453
21	−0.0966	0.4623	33	−0.0541	0.5528
22	−0.0930	0.4700	34	−0.0506	0.5601
23	−0.0895	0.4776	35	−0.0470	0.5675
24	−0.0859	0.4853	36	−0.0435	0.5748
25	−0.0824	0.4928	37	−0.0400	0.5822

9. a)

a	x	k				
		−0.3	−0.1	0.1	0.3	0.5
1.00	0.15	0.1343	0.1350	0.1357	0.1363	0.1370
	0.35	0.2504	0.2540	0.2577	0.2615	0.2653
	0.55	0.3146	0.3221	0.3297	0.3375	0.3456
	0.75	0.3445	0.3558	0.3676	0.3799	0.3927
	1.00	0.3529	0.3691	0.3862	0.4044	0.4235

Continuation of 9. a)

a	x	k				
		−0.3	−0.1	0.1	0.3	0.5
1.25	0.40	0.2567	0.2611	0.2656	0.2702	0.2749
	0.70	0.3062	0.3161	0.3265	0.3372	0.3484
	1.00	0.3060	0.3216	0.3381	0.3557	0.3743
1.50	1.00	0.2702	0.2854	0.3015	0.3186	0.3368
1.75	1.00	0.2471	0.2620	0.2778	0.2947	0.3126
2.00	1.00	0.2379	0.2527	0.2685	0.2852	0.3031

9. b)

a	x		a	x	
	0.6	1.0		0.6	1.0
14	−0.2082	0.3125	26	−0.1742	0.4031
15	−0.2053	0.3201	27	−0.1714	0.4105
16	−0.2025	0.3277	28	−0.1686	0.4180
17	−0.1997	0.3353	29	−0.1657	0.4254
18	−0.1968	0.3429	30	−0.1629	0.4328
19	−0.1940	0.3505	31	−0.1601	0.4402
20	−0.1912	0.3580	32	−0.1573	0.4476
21	−0.1884	0.3656	33	−0.1545	0.4550
22	−0.1855	0.3731	34	−0.1516	0.4623
23	−0.1827	0.3806	35	−0.1488	0.4696
24	−0.1799	0.3881	36	−0.1460	0.4770
25	−0.1771	0.3956	37	−0.1432	0.4843

10. a)

a	x	k				
		1.00	1.25	1.50	1.75	2.00
0.2	0.2	0.1980	0.1974	0.1968	0.1962	0.1956
	0.4	0.3839	0.3792	0.3746	0.3702	0.3658
	0.6	0.5486	0.5347	0.5216	0.5092	0.4975
	0.8	0.6872	0.6596	0.6346	0.6118	0.5910
	1.0	0.7990	0.7554	0.7173	0.6837	0.6538
0.4	0.4	0.3878	0.3830	0.3784	0.3738	0.3694
	0.7	0.6389	0.6176	0.5979	0.5796	0.5627
	1.0	0.8390	0.7915	0.7501	0.7138	0.6816
0.6	0.5	0.4839	0.4748	0.4661	0.4577	0.4496
	1.0	0.8811	0.8294	0.7846	0.7453	0.7106
0.8	1.0	0.9256	0.8693	0.8207	0.7783	0.7409
1.0	1.0	0.9724	0.9113	0.8586	0.8128	0.7725

10. b)

a	x	k 0.50	0.75	1.00	1.25	1.50
1.00	0.2	0.1952	0.1952	0.1952	0.1953	0.1953
	0.4	0.3630	0.3638	0.3645	0.3652	0.3660
	0.6	0.4887	0.4924	0.4958	0.4991	0.5022
	0.8	0.5709	0.5807	0.5898	0.5982	0.6062
	1.0	0.6158	0.6356	0.6538	0.6706	0.6860
1.25	0.4	0.3538	0.3548	0.3557	0.3566	0.3575
	0.7	0.4985	0.5061	0.5132	0.5200	0.5263
	1.0	0.5407	0.5647	0.5867	0.6070	0.6256
1.50	0.5	0.3973	0.4001	0.4028	0.4053	0.4078
	1.0	0.4647	0.4932	0.5193	0.5432	0.5653
1.75	1.0	0.3891	0.4223	0.4527	0.4806	0.5062
2.00	1.0	0.3151	0.3532	0.3881	0.4201	0.4496

10. c)

a	x	k −0.3	−0.1	0.1	0.3	0.5
1.00	0.2	0.1599	0.1615	0.1631	0.1648	0.1664
	0.4	0.2561	0.2609	0.2658	0.2708	0.2759
	0.6	0.3079	0.3163	0.3250	0.3340	0.3434
	0.8	0.3304	0.3424	0.3550	0.3683	0.3822
	1.0	0.3349	0.3506	0.3672	0.3849	0.4037
1.25	0.4	0.2407	0.2453	0.2501	0.2549	0.2599
	0.7	0.2871	0.2969	0.3070	0.3177	0.3288
	1.0	0.2866	0.3016	0.3176	0.3345	0.3526
1.50	0.5	0.2438	0.2499	0.2561	0.2626	0.2693
	1.0	0.2501	0.2646	0.2800	0.2964	0.3140
1.75	1.0	0.2270	0.2411	0.2561	0.2723	0.2895
2.00	1.0	0.2188	0.2328	0.2477	0.2637	0.2808

10. d)

a	x	k −0.5	−0.3	−0.1	0.1	0.3
1.00	0.2	0.1801	0.1806	0.1810	0.1814	0.1819
	0.4	0.3225	0.3255	0.3285	0.3317	0.3349
	0.6	0.4318	0.4401	0.4488	0.4579	0.4675
	0.8	0.5138	0.5298	0.5471	0.5658	0.5860
	1.0	0.5739	0.5994	0.6277	0.6594	0.6951

Continuation of 10. d)

a	x	k = −0.5	−0.3	−0.1	0.1	0.3
1.25	0.4	0.2688	0.2709	0.2730	0.2751	0.2773
	0.7	0.4082	0.4168	0.4258	0.4354	0.4454
	1.0	0.5045	0.5238	0.5450	0.5683	0.5941
1.50	0.5	0.2768	0.2795	0.2823	0.2852	0.2881
	1.0	0.4496	0.4647	0.4811	0.4989	0.5184
1.75	1.0	0.4052	0.4173	0.4304	0.4444	0.4596
2.00	1.0	0.3686	0.3785	0.3892	0.4005	0.4126

10. e)

a	x = 0.6	x = 1.0	a	x = 0.6	x = 1.0
14	−0.0363	0.4844	26	0.0138	0.5753
15	−0.0321	0.4921	27	0.0180	0.5827
16	−0.0279	0.4998	28	0.0221	0.5901
17	−0.0237	0.5075	29	0.0262	0.5974
18	−0.0195	0.5152	30	0.0304	0.6047
19	−0.0153	0.5228	31	0.0345	0.6120
20	−0.0111	0.5304	32	0.0386	0.6192
21	−0.0070	0.5380	33	0.0427	0.6264
22	−0.0028	0.5455	34	0.0468	0.6336
23	0.0014	0.5530	35	0.0510	0.6407
24	0.0055	0.5605	36	0.0551	0.6479
25	0.0097	0.5679	37	0.0592	0.6550

11. a)

a	x	k = −0.3	−0.1	0.1	0.3	0.5
1.00	0.15	0.1272	0.1281	0.1291	0.1300	0.1310
	0.35	0.2377	0.2415	0.2455	0.2495	0.2536
	0.55	0.2992	0.3066	0.3142	0.3222	0.3303
	0.75	0.3278	0.3389	0.3505	0.3626	0.3753
	1.00	0.3359	0.3516	0.3682	0.3858	0.4045
1.25	0.40	0.2417	0.2462	0.2509	0.2556	0.2605
	0.70	0.2883	0.2980	0.3081	0.3186	0.3296
	1.00	0.2879	0.3028	0.3187	0.3356	0.3537
1.50	1.00	0.2516	0.2660	0.2814	0.2978	0.3153
1.75	1.00	0.2287	0.2428	0.2579	0.2740	0.2912
2.00	1.00	0.2210	0.2350	0.2499	0.2659	0.2830

11. b)

a	x = 0.6	x = 1.0	a	x = 0.6	x = 1.0
14	−0.0384	0.4851	26	0.0119	0.5765
15	−0.0342	0.4929	27	0.0161	0.5839
16	−0.0300	0.5007	28	0.0202	0.5913
17	−0.0258	0.5084	29	0.0244	0.5986
18	−0.0216	0.5161	30	0.0285	0.6059
19	−0.0174	0.5238	31	0.0327	0.6132
20	−0.0132	0.5314	32	0.0368	0.6205
21	−0.0090	0.5390	33	0.0410	0.6277
22	−0.0048	0.5465	34	0.0451	0.6349
23	−0.0006	0.5541	35	0.0493	0.6421
24	0.0036	0.5616	36	0.0534	0.6492
25	0.0077	0.5690	37	0.0575	0.6563

12. a)

a	x	k = 1.00	1.25	1.50	1.75	2.00
0.2	0.2	0.1976	0.1969	0.1962	0.1954	0.1947
	0.4	0.3832	0.3783	0.3736	0.3689	0.3644
	0.6	0.5477	0.5336	0.5204	0.5079	0.4961
	0.8	0.6863	0.6586	0.6335	0.6107	0.5898
	1.0	0.7981	0.7545	0.7164	0.6828	0.6530
0.4	0.4	0.3873	0.3823	0.3774	0.3727	0.3681
	0.7	0.6381	0.6166	0.5968	0.5785	0.5615
	1.0	0.8381	0.7906	0.7492	0.7129	0.6807
0.6	0.5	0.4835	0.4741	0.4652	0.4566	0.4484
	1.0	0.8804	0.8286	0.7837	0.7444	0.7097
0.8	1.0	0.9249	0.8685	0.8199	0.7774	0.7400
1.0	1.0	0.9718	0.9105	0.8578	0.8119	0.7716

12. b)

a	x	k = 0.50	0.75	1.00	1.25	1.50
1.0	0.2	0.1942	0.1943	0.1943	0.1944	0.1944
	0.4	0.3615	0.3624	0.3632	0.3639	0.3647
	0.6	0.4872	0.4909	0.4945	0.4978	0.5010
	0.8	0.5696	0.5795	0.5887	0.5973	0.6053
	1.0	0.6148	0.6348	0.6531	0.6699	0.6854

Continuation of 12. b)

a	x	k 0.50	0.75	1.00	1.25	1.50
1.25	0.4	0.3520	0.3531	0.3541	0.3550	0.3560
	0.7	0.4969	0.5046	0.5119	0.5187	0.5252
	1.0	0.5397	0.5639	0.5860	0.6064	0.6251
1.50	0.5	0.3952	0.3981	0.4009	0.4036	0.4061
	1.0	0.4637	0.4924	0.5187	0.5428	0.5649
1.75	1.0	0.3882	0.4216	0.4522	0.4803	0.5060
2.00	1.0	0.3143	0.3527	0.3879	0.4201	0.4497

12. c)

a	x	k −0.3	−0.1	0.1	0.3	0.5
1.00	0.2	0.1601	0.1617	0.1633	0.1649	0.1665
	0.4	0.2565	0.2613	0.2662	0.2712	0.2763
	0.6	0.3086	0.3170	0.3257	0.3347	0.3440
	0.8	0.3312	0.3433	0.3559	0.3692	0.3832
	1.0	0.3359	0.3517	0.3684	0.3861	0.4049
1.25	0.4	0.2413	0.2459	0.2507	0.2555	0.2605
	0.7	0.2882	0.2980	0.3082	0.3189	0.3300
	1.0	0.2882	0.3033	0.3193	0.3364	0.3545
1.50	0.5	0.2448	0.2509	0.2572	0.2637	0.2704
	1.0	0.2524	0.2669	0.2825	0.2990	0.3167
1.75	1.0	0.2302	0.2444	0.2597	0.2759	0.2934
2.00	1.0	0.2232	0.2373	0.2525	0.2687	0.2860

12. d)

a	x	k −0.5	−0.3	−0.1	0.1	0.3
1.00	0.2	0.1801	0.1806	0.1812	0.1817	0.1822
	0.4	0.3225	0.3256	0.3288	0.3321	0.3355
	0.6	0.4318	0.4403	0.4492	0.4585	0.4682
	0.8	0.5139	0.5301	0.5475	0.5664	0.5869
	1.0	0.5740	0.5997	0.6282	0.6601	0.6962
1.25	0.4	0.2688	0.2709	0.2731	0.2753	0.2776
	0.7	0.4081	0.4169	0.4261	0.4357	0.4459
	1.0	0.5045	0.5240	0.5453	0.5688	0.5948
1.50	0.5	0.2767	0.2795	0.2824	0.2853	0.2884
	1.0	0.4495	0.4648	0.4813	0.4992	0.5188
1.75	1.0	0.4051	0.4173	0.4305	0.4446	0.4599
2.00	1.0	0.3685	0.3785	0.3892	0.4007	0.4129

12. e)

a	x = 0.6	x = 1.0	a	x = 0.6	x = 1.0
14	−0.0395	0.4790	26	0.0102	0.5695
15	−0.0353	0.4867	27	0.0143	0.5769
16	−0.0312	0.4944	28	0.0184	0.5842
17	−0.0270	0.5021	29	0.0225	0.5915
18	−0.0229	0.5097	30	0.0266	0.5987
19	−0.0187	0.5173	31	0.0307	0.6059
20	−0.0146	0.5248	32	0.0348	0.6131
21	−0.0104	0.5323	33	0.0388	0.6203
22	−0.0063	0.5398	34	0.0429	0.6275
23	−0.0022	0.5473	35	0.0470	0.6346
24	0.0019	0.5547	36	0.0511	0.6417
25	0.0061	0.5621	37	0.0551	0.6487

13. a)

a	x	k = −0.3	k = −0.1	k = 0.1	k = 0.3	k = 0.5
1.00	0.15	0.1272	0.1282	0.1291	0.1300	0.1310
	0.35	0.2378	0.2416	0.2455	0.2495	0.2536
	0.55	0.2993	0.3067	0.3144	0.3223	0.3305
	0.75	0.3280	0.3391	0.3507	0.3628	0.3755
	1.00	0.3362	0.3518	0.3684	0.3861	0.4048
1.25	0.40	0.2418	0.2464	0.2510	0.2558	0.2607
	0.70	0.2885	0.2982	0.3083	0.3189	0.3299
	1.00	0.2882	0.3032	0.3191	0.3361	0.3541
1.50	1.00	0.2521	0.2666	0.2820	0.2984	0.3160
1.75	1.00	0.2295	0.2436	0.2587	0.2749	0.2921
2.00	1.00	0.2221	0.2361	0.2511	0.2671	0.2843

13. b)

a	x = 0.6	x = 1.0	a	x = 0.6	x = 1.0
14	−0.0505	0.4641	26	−0.0017	0.5534
15	−0.0465	0.4717	27	0.0023	0.5606
16	−0.0424	0.4793	28	0.0063	0.5678
17	−0.0383	0.4869	29	0.0104	0.5750
18	−0.0342	0.4944	30	0.0144	0.5822
19	−0.0301	0.5019	31	0.0184	0.5893
20	−0.0261	0.5093	32	0.0224	0.5964
21	−0.0220	0.5167	33	0.0264	0.6034
22	−0.0179	0.5241	34	0.0304	0.6105
23	−0.0139	0.5315	35	0.0344	0.6175
24	−0.0098	0.5388	36	0.0384	0.6245
25	−0.0058	0.5461	37	0.0424	0.6314

14. a)

a	x	k				
		1.00	1.25	1.50	1.75	2.00
0.2	0.2	0.1976	0.1969	0.1962	0.1955	0.1948
	0.4	0.3834	0.3786	0.3739	0.3694	0.3650
	0.6	0.5481	0.5342	0.5212	0.5089	0.4973
	0.8	0.6870	0.6596	0.6348	0.6122	0.5915
	1.0	0.7991	0.7557	0.7178	0.6844	0.6547
0.4	0.4	0.3874	0.3825	0.3777	0.3731	0.3687
	0.7	0.6385	0.6172	0.5977	0.5796	0.5628
	1.0	0.8389	0.7917	0.7505	0.7144	0.6824
0.6	0.5	0.4836	0.4743	0.4655	0.4571	0.4491
	1.0	0.8810	0.8295	0.7849	0.7458	0.7113
0.8	1.0	0.9253	0.8693	0.8209	0.7787	0.7415
1.0	1.0	0.9721	0.9111	0.8587	0.8130	0.7730

14. b)

a	x	k				
		0.50	0.75	1.00	1.25	1.50
1.00	0.2	0.1944	0.1945	0.1946	0.1946	0.1947
	0.4	0.3621	0.3631	0.3640	0.3648	0.3657
	0.6	0.4884	0.4922	0.4959	0.4993	0.5025
	0.8	0.5712	0.5812	0.5904	0.5990	0.6070
	1.0	0.6165	0.6365	0.6548	0.6716	0.6870
1.25	0.4	0.3528	0.3540	0.3551	0.3562	0.3572
	0.7	0.4986	0.5064	0.5138	0.5207	0.5272
	1.0	0.5417	0.5659	0.5880	0.6082	0.6268
1.50	0.5	0.3966	0.3997	0.4026	0.4054	0.4080
	1.0	0.4660	0.4946	0.5208	0.5447	0.5667
1.75	1.0	0.3906	0.4240	0.4544	0.4823	0.5079
2.00	1.0	0.3169	0.3552	0.3901	0.4222	0.4515

14. c)

a	x	k				
		−0.3	−0.1	0.1	0.3	0.5
1.00	0.2	0.1611	0.1625	0.1640	0.1656	0.1671
	0.4	0.2579	0.2626	0.2673	0.2722	0.2772
	0.6	0.3101	0.3184	0.3269	0.3358	0.3450
	0.8	0.3327	0.3447	0.3572	0.3704	0.3841
	1.0	0.3374	0.3530	0.3696	0.3872	0.4059
1.25	0.4	0.2428	0.2473	0.2519	0.2566	0.2614
	0.7	0.2898	0.2995	0.3095	0.3200	0.3310
	1.0	0.2897	0.3047	0.3206	0.3375	0.3555

Continuation of 14. c)

a	x	k: −0.3	−0.1	0.1	0.3	0.5
1.50	0.5	0.2465	0.2524	0.2586	0.2649	0.2714
	1.0	0.2540	0.2684	0.2838	0.3003	0.3178
1.75	1.0	0.2319	0.2461	0.2612	0.2774	0.2946
2.00	1.0	0.2252	0.2393	0.2543	0.2704	0.2877

14. d)

a	x	k: −0.5	−0.3	−0.1	0.1	0.3
1.00	0.2	0.1802	0.1807	0.1812	0.1817	0.1822
	0.4	0.3228	0.3259	0.3290	0.3322	0.3355
	0.6	0.4324	0.4407	0.4495	0.4586	0.4683
	0.8	0.5146	0.5306	0.5479	0.5666	0.5869
	1.0	0.5748	0.6003	0.6286	0.6603	0.6962
1.25	0.4	0.2690	0.2711	0.2732	0.2754	0.2776
	0.7	0.4085	0.4171	0.4262	0.4358	0.4459
	1.0	0.5050	0.5243	0.5455	0.5689	0.5948
1.50	0.5	0.2769	0.2796	0.2825	0.2854	0.2884
	1.0	0.4499	0.4650	0.4814	0.4993	0.5188
1.75	1.0	0.4053	0.4175	0.4306	0.4446	0.4599
2.00	1.0	0.3687	0.3786	0.3893	0.4007	0.4129

14. e)

a	x: 0.6	x: 1.0	a	x: 0.6	x: 1.0
14	−0.0370	0.4814	26	0.0128	0.5721
15	−0.0328	0.4891	27	0.0169	0.5794
16	−0.0286	0.4968	28	0.0210	0.5867
17	−0.0245	0.5045	29	0.0251	0.5940
18	−0.0203	0.5121	30	0.0292	0.6013
19	−0.0162	0.5197	31	0.0333	0.6085
20	−0.0120	0.5273	32	0.0374	0.6158
21	−0.0079	0.5348	33	0.0415	0.6229
22	−0.0037	0.5423	34	0.0456	0.6301
23	0.0004	0.5498	35	0.0496	0.6372
24	0.0045	0.5572	36	0.0537	0.6443
25	0.0087	0.5647	37	0.0578	0.6514

15. a)

a	x	k				
		−0.3	−0.1	0.1	0.3	0.5
1.00	0.15	0.1274	0.1283	0.1292	0.1302	0.1311
	0.35	0.2381	0.2419	0.2458	0.2498	0.2538
	0.55	0.2997	0.3071	0.3147	0.3226	0.3307
	0.75	0.3284	0.3395	0.3510	0.3631	0.3757
	0.95	0.3366	0.3513	0.3668	0.3833	0.4007
1.25	0.35	0.2260	0.2297	0.2334	0.2373	0.2412
	0.65	0.2858	0.2946	0.3037	0.3132	0.3231
	0.95	0.2900	0.3040	0.3189	0.3346	0.3514
1.50	0.45	0.2379	0.2430	0.2483	0.2538	0.2594
	0.95	0.2532	0.2667	0.2811	0.2963	0.3125
1.75	0.95	0.2278	0.2409	0.2548	0.2697	0.2855
2.00	0.95	0.2150	0.2279	0.2416	0.2562	0.2719

15. b)

a	x		a	x	
	0.6	1.0		0.6	1.0
14	−0.0412	0.4730	26	0.0078	0.5629
15	−0.0370	0.4807	27	0.0118	0.5702
16	−0.0329	0.4883	28	0.0159	0.5774
17	−0.0288	0.4959	29	0.0199	0.5847
18	−0.0248	0.5035	30	0.0240	0.5919
19	−0.0207	0.5110	31	0.0280	0.5991
20	−0.0166	0.5185	32	0.0320	0.6062
21	−0.0125	0.5260	33	0.0360	0.6133
22	−0.0084	0.5334	34	0.0400	0.6205
23	−0.0044	0.5408	35	0.0440	0.6275
24	−0.0003	0.5482	36	0.0480	0.6346
25	0.0038	0.5555	37	0.0520	0.6416

16. a)

a	x	k				
		1.00	1.25	1.50	1.75	2.00
0.50	0.2	0.07663	0.07586	0 07531	0.07469	0.07408
	0.4	0.11223	0.11086	0.10956	0.10831	0.10712
	0.6	0.11455	0.11308	0.11169	0.11035	0.10908
	0.8	0.09119	0.09003	0.08893	0.08788	0.08687
	1.0	0.04952	0.04870	0.04793	0.04720	0.04650
0.75	0.4	0.19861	0.19467	0.19101	0.18759	0.18439
	0.7	0.24925	0.24324	0.23773	0.23266	0.22796
	1.0	0.24457	0.23819	0.23238	0.22706	0.22216

Continuation of 16. a)

a	x	k				
		1.00	1.25	1.50	1.75	2.00
1.00	0.5	0.32346	0.31400	0.30548	0.29775	0.29069
	1.0	0.43110	0.41478	0.40047	0.38778	0.37641
1.25	1.0	0.61007	0.58083	0.55598	0.53449	0.51563
1.50	1.0	0.78226	0.73804	0.70141	0.67035	0.64354

16. b)

a	x	k				
		10.0	12.5	15.0	17.5	20.0
1.00	0.2	0.05096	0.05085	0.05075	0.05064	0.05054
	0.4	0.10263	0.10181	0.10101	0.10024	0.09950
	0.6	0.15333	0.15079	0.14843	0.14623	0.14416
	0.7	0.20183	0.19658	0.19187	0.18760	0.18370
	1.0	0.24753	0.23876	0.23114	0.22442	0.21843
1.25	0.4	0.10389	0.10303	0.10221	0.10142	0.10065
	0.7	0.18144	0.17748	0.17386	0.17054	0.16748
	1.0	0.25393	0.24463	0.23658	0.22950	0.22320
1.50	0.5	0.13206	0.13039	0.12881	0.12731	0.12589
	1.0	0.26051	0.25066	0.24216	0.23470	0.22809
1.75	1.0	0.26727	0.25684	0.24786	0.24002	0.23307
2.00	1.0	0.27421	0.26317	0.25371	0.24545	0.23816

16. c)

a	x	k				
		10.0	12.5	15.0	17.5	20.0
1.0	0.2	0.05006	0.05004	0.05001	0.04998	0.04996
	0.4	0.10049	0.10029	0.10008	0.09988	0.09968
	0.6	0.15162	0.15094	0.15026	0.14961	0.14897
	0.8	0.20374	0.20213	0.20060	0.19912	0.19771
	1.0	0.25704	0.25397	0.25110	0.24840	0.24585
1.5	0.4	0.10115	0.10094	0.10073	0.10052	0.10032
	0.7	0.18100	0.17984	0.17872	0.17764	0.17660
	1.0	0.26672	0.26327	0.26005	0.25704	0.25421
2.0	0.5	0.12853	0.12810	0.12767	0.12726	0.12685
	1.0	0.27674	0.27286	0.26927	0.26592	0.26278
2.5	1.0	0.28709	0.28275	0.27874	0.27502	0.27155
3.0	1.0	0.29777	0.29292	0.28847	0.28436	0.28053

16. d)

a	x	k 10.0	12.5	15.0	17.5	20.0
1.0	0.2	0.03924	0.03888	0.03853	0.03820	0.03788
	0.4	0.07711	0.07581	0.07458	0.07343	0.07234
	0.6	0.11381	0.11112	0.10865	0.10638	0.10428
	0.8	0.14950	0.14508	0.14111	0.13752	0.13424
	1.0	0.18432	0.17790	0.17223	0.16718	0.16263
2.0	0.4	0.07712	0.07581	0.07459	0.07344	0.07235
	0.7	0.13184	0.12831	0.12512	0.12220	0.11953
	1.0	0.18460	0.17813	0.17244	0.16736	0.16279
3.0	0.5	0.09564	0.09368	0.09187	0.09019	0.08862
	1.0	0.18487	0.17837	0.17264	0.16754	0.16295
4.0	1.0	0.18514	0.17860	0.17285	0.16772	0.16311
5.0	1.0	0.18542	0.17884	0.17305	0.16790	0.16327

17. a)

a	x	k 1.00	1.25	1.50	1.75	2.00
0.50	0.15	0.06165	0.06121	0.06078	0.06036	0.05996
	0.35	0.10676	0.10552	0.10434	0.10320	0.10211
	0.55	0.11669	0.11521	0.11379	0.11243	0.11114
	0.75	0.09905	0.09780	0.09661	0.09548	0.09439
	0.95	0.06130	0.06042	0.05958	0.05879	0.05803
0.75	0.35	0.18318	0.17978	0.17661	0.17364	0.17085
	0.65	0.24528	0.23948	0.23416	0.22925	0.22470
	0.95	0.24834	0.24195	0.23612	0.23078	0.22586
1.00	0.45	0.30325	0.29481	0.28718	0.28024	0.27386
	0.95	0.42676	0.41088	0.39693	0.38454	0.37342
1.25	0.95	0.59762	0.56956	0.54565	0.52493	0.50671
1.50	0.95	0.76177	0.71969	0.68471	0.65497	0.62924

17. b)

a	x	k 10.0	12.5	15.0	17.5	20.0
1.0	0.15	0.02957	0.02936	0.02916	0.02896	0.02877
	0.35	0.06776	0.06674	0.06577	0.06486	0.06399
	0.55	0.10473	0.10242	0.10030	0.09833	0.09650
	0.75	0.14066	0.13670	0.13313	0.12988	0.12691
	0.95	0.17569	0.16979	0.16457	0.15989	0.15566

Continuation of 17. b)

a	x	k 10.0	12.5	15.0	17.5	20.0
2.0	0.35	0.06777	0.06674	0.06578	0.06486	0.06400
	0.65	0.12287	0.11977	0.11695	0.11436	0.11197
	0.95	0.17592	0.16998	0.16474	0.16004	0.15580
3.0	0.45	0.08642	0.08480	0.08329	0.08188	0.08056
	0.95	0.17614	0.17018	0.16491	0.16019	0.15593
4.0	0.95	0.17637	0.17037	0.16508	0.16034	0.15607
5.0	0.95	0.17659	0.17057	0.16525	0.16049	0.15620

18. a)

a	x	k 1.00	1.25	1.50	1.75	2.00
0.4	0.2	0.07560	0.06085	0.05089	0.04372	0.03832
	0.4	0.12846	0.10477	0.08836	0.07634	0.06718
	0.6	0.14811	0.12219	0.10382	0.09017	0.07965
	0.8	0.13430	0.11080	0.09408	0.08164	0.07204
	1.0	0.09019	0.07183	0.05917	0.05003	0.04317
0.8	0.4	0.26680	0.22004	0.18681	0.16211	0.14307
	0.7	0.34814	0.29557	0.25598	0.22531	0.20096
	1.0	0.34208	0.29435	0.25740	0.22819	0.20463
1.2	0.5	0.45540	0.38433	0.33125	0.29044	0.25825
	1.0	0.55294	0.48755	0.43426	0.39044	0.35401
1.6	1.0	0.73662	0.65975	0.59487	0.54003	0.49341
2.0	1.0	0.90075	0.81605	0.74265	0.67925	0.62442

18. b)

a	x	k 10.0	12.5	15.0	17.5	20.0
1.0	0.2	0.08881	0.08692	0.08517	0.08355	0.08203
	0.4	0.16463	0.15919	0.15436	0.15002	0.14609
	0.6	0.23523	0.22556	0.21718	0.20981	0.20325
	0.8	0.30460	0.29018	0.27791	0.26729	0.25795
	1.0	0.37519	0.35549	0.33898	0.32484	0.31254
1.5	0.4	0.16496	0.15949	0.15463	0.15026	0.14631
	0.7	0.27119	0.25903	0.24861	0.23953	0.23151
	1.0	0.37822	0.35804	0.34117	0.32676	0.31424
2.0	0.5	0.20144	0.19381	0.18714	0.18122	0.17593
	1.0	0.38131	0.36063	0.34339	0.32870	0.31596
2.5	1.0	0.38445	0.36326	0.34565	0.33067	0.31769
3.0	1.0	0.38765	0.36593	0.34793	0.33266	0.31945

18. c)

a	x	k = 10.0	12.5	15.0	17.5	20.0
1.0	0.2	0.04870	0.04868	0.04866	0.04863	0.04861
	0.4	0.09474	0.09457	0.09441	0.09426	0.09410
	0.6	0.13818	0.13771	0.13724	0.13679	0.13634
	0.8	0.17929	0.17831	0.17737	0.17645	0.17557
	1.0	0.21814	0.21675	0.21516	0.21363	0.21216
2.0	0.4	0.09482	0.09466	0.09450	0.09434	0.09418
	0.7	0.15968	0.15897	0.15828	0.15760	0.15694
	1.0	0.22087	0.21913	0.21746	0.21586	0.21433
3.0	0.5	0.11716	0.11686	0.11657	0.11628	0.11599
	1.0	0.22338	0.22155	0.21981	0.21815	0.21655
4.0	1.0	0.22595	0.22404	0.22222	0.22048	0.21881
5.0	1.0	0.22858	0.22659	0.22468	0.22287	0.22113

18. d)

a	x	k = 1.0	1.5	2.0	2.5	3.0
1.0	0.2	0.19729	0.13154	0.09866	0.07893	0.06578
	0.4	0.37723	0.25204	0.18917	0.15138	0.12617
	0.6	0.51648	0.34895	0.26276	0.21058	0.17564
	0.8	0.57905	0.40712	0.30971	0.24926	0.20836
	1.0	0.52507	0.40948	0.31953	0.25981	0.21832
1.5	0.4	0.36589	0.24474	0.18375	0.14707	0.12259
	0.7	0.49430	0.34335	0.26061	0.20957	0.17512
	1.0	0.35069	0.29196	0.23313	0.19151	0.16180
2.0	0.5	0.40839	0.27549	0.20739	0.16619	0.13861
	1.0	0.19257	0.17993	0.14893	0.12430	0.10590
2.5	1.0	0.04597	0.07251	0.06675	0.05811	0.05059
3.0	1.0	−0.09183	−0.03096	−0.01361	−0.00710	−0.00415

19. a)

a	x	k = 10.0	12.5	15.0	17.5	20.0
1.0	0.15	0.068273	0.067092	0.065986	0.064946	0.063966
	0.35	0.146378	0.141904	0.137895	0.134272	0.130973
	0.55	0.217814	0.209252	0.201790	0.195199	0.189313
	0.75	0.287221	0.274037	0.262773	0.252983	0.244356
	0.95	0.357331	0.339002	0.323588	0.310357	0.298818

Continuation of 19. a)

a	x	k				
		10.0	12.5	15.0	17.5	20.0
1.5	0.35	0.146614	0.142117	0.138089	0.134450	0.131136
	0.65	0.253659	0.242704	0.233271	0.225018	0.217707
	0.95	0.360017	0.341271	0.325544	0.312071	0.300338
2.0	0.45	0.183466	0.176921	0.171153	0.166011	0.161382
	0.95	0.362748	0.343573	0.327526	0.313804	0.301874
2.5	0.95	0.365524	0.345908	0.329532	0.315558	0.303426
3.0	0.95	0.368345	0.348277	0.331565	0.317332	0.304995

19. b)

a	x	k				
		1.0	1.5	2.0	2.5	3.0
1.0	0.15	0.148865	0.099247	0.074436	0.059549	0.049625
	0.35	0.334980	0.223598	0.167768	0.134239	0.111877
	0.55	0.487187	0.327716	0.246462	0.197408	0.164613
	0.75	0.573069	0.397139	0.301039	0.241925	0.202082
	0.95	0.550397	0.414998	0.321122	0.260225	0.218294
1.5	0.35	0.327488	0.218735	0.164154	0.131360	0.109483
	0.65	0.488786	0.335458	0.253767	0.203777	0.170152
	0.95	0.395158	0.313464	0.246834	0.201580	0.169793
2.0	0.45	0.384329	0.258124	0.194056	0.15511	0.129588
	0.95	0.253380	0.215935	0.174116	0.143706	0.121724
2.5	0.95	0.121325	0.121854	0.102848	0.086566	0.074083
3.0	0.95	−0.003205	0.030789	0.032924	0.030124	0.026846

20. a)

a	k				
	1.0	1.25	1.50	1.75	2.0
0.5	0.31958	0.31111	0.30333	0.29616	0.28952
0.7	0.42696	0.41675	0.40730	0.39853	0.39034
0.9	0.51945	0.50845	0.49818	0.48855	0.47950
1.1	0.59770	0.58660	0.57614	0.56626	0.55691
1.3	0.66324	0.65248	0.64226	0.63254	0.62329

20. b)

a	k				
	0	0.5	1.0	1.5	2.0
−2.0	−0.92603	−0.87781	−0.83207	−0.78943	−0.75008
−1.5	−0.73295	−0.68718	−0.64484	−0.60648	−0.57203
−1.0	−0.48441	−0.44989	−0.41844	−0.39073	−0.36648
−0.5	−0.13813	−0.14255	−0.13968	−0.13475	−0.12933
0	0.33680	0.24673	0.19510	0.16166	0.13820

20. c)

a	0.50	0.75	1.00	1.25	1.50
y (1)	0.00910	0.31154	0.68695	1.17119	1.82858

20. d)

a	14	15	16	17	18
y (1)	0.48420	0.49195	0.49967	0.50735	0.51500
a	19	20	21	22	23
y (1)	0.52262	0.53021	0.53777	0.54530	0.55279
a	24	25	26	27	28
y (1)	0.56026	0.56770	0.57510	0.58248	0.58983
a	29	30	31	32	33
y (1)	0.59715	0.60444	0.61170	0.61893	0.62614
a	34	35	36	37	38
y (1)	0.63331	0.64046	0.64758	0.65468	0.66174

21. a)

a	x	k			
		1	2	3	4
1.0	0.05	−0.00119	−0.00004	0.00000	0.00000
	0.10	−0.00455	−0.00033	−0.00002	0.00000
	0.15	−0.00982	−0.00110	−0.00013	−0.00002
	0.20	−0.01678	−0.00257	−0.00040	−0.00006
1.3	0.05	−0.00153	−0.00005	0.00000	0.00000
	0.10	−0.00577	−0.00043	−0.00003	0.00000
	0.15	−0.01232	−0.00142	−0.00016	−0.00002
	0.20	−0.02086	−0.00330	−0.00051	−0.00008
	0.25	−0.03114			
1.6	0.05	−0.00185	−0.00007	0.00000	0.00000
	0.10	−0.00693	−0.00052	−0.00004	0.00000
	0.15	−0.01465	−0.00174	−0.00020	−0.00002
	0.20	−0.02461	−0.00401	−0.00063	−0.00010
	0.25	−0.03647	−0.00760		

21. b)

a	k = 0.05	0.10	0.15	0.20
0.2	0.99975	0.99900		
0.4	0.99950	0.99800	0.99551	
0.6	0.99925	0.99700	0.99327	
0.8	0.99900	0.99601	0.99104	
1.0	0.99875	0.99501	0.98881	
1.2	0.99850	0.99402	0.98659	
1.4	0.99825	0.99302	0.98437	
1.6	0.99800	0.99203	0.98216	
1.8	0.99775	0.99104	0.97995	0.96464
2.0	0.99750	0.99005	0.97775	0.96079

21. c)

a	x	k = 2.5	2.6	2.7	2.8	2.9
2.0	2.05	1.95947	1.96116	1.96271	1.96413	1.96544
	2.10	1.92074	1.92401	1.92700	1.92976	1.93231
	2.15	1.88368	1.88842	1.89278	1.89679	1.90051
2.2	2.05	1.95567	1.95750	1.95917	1.96071	1.96214
	2.10	1.91339	1.91691	1.92014	1.92313	1.92589
	2.15	1.87301	1.87811	1.88280	1.88714	1.89115
2.4	2.05	1.95189	1.95384	1.95564	1.95730	1.95884
	2.10	1.90608	1.90985	1.91332	1.91653	1.91950
	2.15	1.86241	1.86787	1.87290	1.87755	1.88186
2.6	2.05	1.94811	1.95020	1.95213	1.95390	1.95555
	2.10	1.89880	1.90282	1.90653	1.90995	1.91313
	2.15	1.85188	1.85769	1.86305	1.86801	1.87262
2.8	2.05	1.94435	1.94657	1.94862	1.95051	1.95226
	2.10	1.89156	1.89582	1.89976	1.90341	1.90679
	2.15	1.84142	1.84757	1.85326	1.85853	1.86342

22. a)

a	x	k = 1	2	3	4
1.0	0.30	−0.03509	−0.00827	−0.00197	−0.00048
	0.60	−0.11676	−0.05419	−0.02696	−0.01384
	0.70	−0.15080	−0.07959	−0.04579	−0.02751
1.3	0.30	−0.04295	−0.01051	−0.00255	−0.00062
	0.60	−0.13846	−0.06598	−0.03348	−0.01745
	0.70	−0.17747	−0.09565	−0.05593	−0.03405
1.6	0.30	−0.05000	−0.01265	−0.00311	−0.00076
	0.60	−0.15700	−0.07651	−0.03948	−0.02085
	0.70	−0.20000	−0.10970	−0.06502	−0.04005

22. b)

a	x	k = 2.5	2.6	2.7	2.8	2.9
2.0	2.30	1.78150	1.79014	1.79811	1.80546	1.81228
	2.55	1.63608	1.64983	1.66254	1.67433	1.68531
	2.60	1.61011	1.62471	1.63823	1.65076	1.66244
	2.65	1.58505	1.60045	1.61473	1.62797	1.64032
	2.70	1.56084	1.57701	1.59200	1.60592	1.61891
2.2	2.30	1.76205	1.77131	1.77984	1.78775	1.79509
	2.55	1.60518	1.61978	1.63332	1.64590	1.65762
	2.60	1.57729	1.59279	1.60715	1.62052	1.63297
	2.65	1.55042	1.56675	1.58190	1.59600	1.60915
	2.70	1.52451	1.54161	1.55751	1.57231	1.58613
2.4	2.30	1.74284	1.75269	1.76180	1.77024	1.77808
	2.55	1.57489	1.59032	1.60465	1.61798	1.63042
	2.60	1.54518	1.56152	1.57671	1.59085	1.60406
	2.65	1.51659	1.53379	1.54978	1.56469	1.57862
	2.70	1.48906	1.50706	1.52381	1.53943	1.55405
2.6	2.30	1.72385	1.73428	1.74394	1.75290	1.76124
	2.55	1.54519	1.56140	1.57648	1.59054	1.60368
	2.60	1.51373	1.53088	1.54684	1.56174	1.57566
	2.65	1.48351	1.50153	1.51832	1.53399	1.54866
	2.70	1.45445	1.47328	1.49084	1.50725	1.52262
2.8	2.30	1.70508	1.71607	1.72627	1.73575	1.74457
	2.55	1.51605	1.53302	1.54882	1.56358	1.57738
	2.60	1.48293	1.50085	1.51755	1.53316	1.54778
	2.65	1.45115	1.46995	1.48750	1.50391	1.51928
	2.70	1.42064	1.44026	1.45859	1.47575	1.49183

23. a)

a	x	k = 1.00	1.25	1.50	1.75	2.00
0.50	1.2	−0.0039	−0.0046	−0.0053	−0.0059	−0.0065
	1.4	−0.0639	−0.0648	−0.0657	−0.0665	−0.0673
	1.6	−0.1266	−0.1280	−0.1294	−0.1307	−0.1321
	1.8	−0.1895	−0.1916	−0.1937	−0.1958	−0.1979
	2.0	−0.2511	−0.2539	−0.2568	−0.2598	−0.2628
0.75	1.4	0.1857	0.1799	0.1747	0.1700	0.1656
	1.7	0.1206	0.1153	0.1105	0.1061	0.1021
	2.0	0.0481	0.0429	0.0383	0.0341	0.0302
1.00	1.5	0.4238	0.4051	0.3890	0.3749	0.3623
	2.0	0.3586	0.3394	0.3229	0.3086	0.2960
1.25	2.0	0.6771	0.6331	0.5971	0.5668	0.5409
1.50	2.0	1.0009	0.9229	0.8614	0.8112	0.7692

23. b)

a	x	k				
		10.0	12.5	15.0	17.5	20.0
1.0	1.2	0.44858	0.42302	0.40186	0.38393	0.36845
	1.4	0.52586	0.49379	0.46755	0.44549	0.42658
	1.6	0.60778	0.56851	0.53669	0.51015	0.48753
	1.8	0.69485	0.64764	0.60973	0.57832	0.55170
	2.0	0.78743	0.73151	0.68695	0.65028	0.61934
1.5	1.4	0.53258	0.49930	0.47219	0.44948	0.43007
	1.7	0.66133	0.61612	0.57987	0.54989	0.52449
	2.0	0.80328	0.74412	0.69733	0.65904	0.62689
2.0	1.5	0.58228	0.54371	0.51261	0.48678	0.46484
	2.0	0.81983	0.75720	0.70806	0.66806	0.63463
2.5	2.0	0.83714	0.77078	0.71913	0.67735	0.64258
3.0	2.0	0.85524	0.78489	0.73058	0.68691	0.65074

23. c)

a	x	k				
		10.0	12.5	15.0	17.5	20.0
1.0	1.2	0.25593	0.25340	0.25100	0.24872	0.24656
	1.4	0.29223	0.28865	0.28530	0.28215	0.27919
	1.6	0.32766	0.32286	0.31842	0.31429	0.31043
	1.8	0.36257	0.35637	0.35069	0.34547	0.34062
	2.0	0.39729	0.38948	0.38242	0.37598	0.37007
2.0	1.4	0.30020	0.29623	0.29252	0.28906	0.28581
	1.7	0.36074	0.35427	0.34838	0.34296	0.33796
	2.0	0.42454	0.41462	0.40579	0.39784	0.39062
3.0	1.5	0.33095	0.32561	0.32070	0.31616	0.31193
	2.0	0.45543	0.44280	0.43175	0.42194	0.41313
4.0	2.0	0.49052	0.47442	0.46059	0.44849	0.43777
5.0	2.0	0.53044	0.50991	0.49262	0.47774	0.46472

23. d)

a	x				
	0.25	0.50	0.55	0.60	0.65
0.2	0.99377	0.97531	0.97020	0.96464	0.95863
0.4	0.98758	0.95123	0.94129	0.93053	0.91897
0.6	0.98142	0.92774	0.91325	0.89763	0.88095
0.8	0.97531	0.90484	0.88603	0.86589	0.84451
1.0	0.96923	0.88250	0.85963	0.83527	0.80957
1.2	0.96319	0.86071	0.83402	0.80574	0.77608
1.4	0.95719	0.83946	0.80917	0.77724	0.74397
1.6	0.95123	0.81873	0.78506	0.74976	0.71320
1.8	0.94530	0.79852	0.76166	0.72325	0.68369
2.0	0.93941	0.77880	0.73897	0.69768	0.65541

24. a)

a	x	k				
		10.0	12.5	15.0	17.5	20.0
1.0	0.5	0.14063	0.14008	0.13954	0.13901	0.13849
	0.8	0.24839	0.24570	0.24316	0.24076	0.23848
	1.0	0.33073	0.32513	0.32000	0.31527	0.31089
1.5	0.5	0.14482	0.14421	0.14361	0.14303	0.14246
	0.8	0.25880	0.25575	0.25288	0.25018	0.24762
	1.0	0.34604	0.33965	0.33384	0.32852	0.32361
2.0	0.5	0.14916	0.14849	0.14783	0.14719	0.14657
	0.8	0.26971	0.26624	0.26301	0.25997	0.25712
	1.0	0.36208	0.35481	0.34825	0.34226	0.33677
2.5	0.5	0.15366	0.15292	0.15220	0.15150	0.15081
	1.0	0.37885	0.37060	0.36320	0.35650	0.35037
3.0	0.5	0.15834	0.15752	0.15672	0.15595	0.15520
	1.0	0.39635	0.38702	0.37870	0.37121	0.36440

24. b)

a	x			
	1.3	1.5	1.8	2.0
0.50	0.6761	0.5017	0.3285	0.2550
0.75	0.7046	0.5468	0.3803	0.3048
1.00	0.7197	0.5725	0.4129	0.3377
1.25	0.7291	0.5892	0.4353	0.3612
1.50	0.7355	0.6008	0.4516	0.3787
1.75	0.7401	0.6094	0.4640	0.3923
2.00	0.7436	0.6160	0.4737	0.4032
2.25	0.7464	0.6213	0.4816	0.4120
2.50	0.7486	0.6255	0.4881	0.4194
2.75	0.7504	0.6291	0.4935	0.4256
3.00	0.7520	0.6320	0.4982	0.4310
3.25	0.7533	0.6346	0.5022	0.4356
3.50	0.7544	0.6368	0.5056	0.4396
3.75	0.7554	0.6387	0.5087	0.4432
4.00	0.7562	0.6403	0.5114	0.4464
4.25	0.7570	0.6418	0.5138	0.4492
4.50	0.7576	0.6432	0.5159	0.4518
4.75	0.7582	0.6444	0.5179	0.4541
5.00	0.7588	0.6454	0.5196	0.4562
5.25	0.7593	0.6464	0.5213	0.4581
5.50	0.7597	0.6473	0.5227	0.4598
5.75	0.7601	0.6481	0.5241	0.4614
6.00	0.7605	0.6489	0.5253	0.4629
6.25	0.7609	0.6496	0.5265	0.4643
6.50	0.7612	0.6502	0.5275	0.4656

24. c)

a	x			
	0.3	0.5	0.8	1.0
0.20	2.4329	3.2747	4.2835	4.7477
0.21	2.3626	3.1606	4.1127	4.5473
0.22	2.2986	3.0568	3.9574	4.3652
0.23	2.2402	2.9620	3.8156	4.1989
0.24	2.1867	2.8752	3.6857	4.0464
0.25	2.1374	2.7953	3.5661	3.9062
0.26	2.0920	2.7215	3.4558	3.7767
0.27	2.0499	2.6532	3.3536	3.6569
0.28	2.0108	2.5898	3.2587	3.5456
0.29	1.9744	2.5308	3.1703	3.4419
0.30	1.9404	2.4757	3.0879	3.3452
0.31	1.9086	2.4241	3.0108	3.2547
0.32	1.8788	2.3758	2.9384	3.1699
0.33	1.8509	2.3304	2.8705	3.0902
0.34	1.8245	2.2877	2.8066	3.0152
0.35	1.7997	2.2474	2.7463	2.9445
0.36	1.7762	2.2093	2.6894	2.8777
0.37	1.7540	2.1733	2.6355	2.8145
0.38	1.7330	2.1392	2.5845	2.7547
0.39	1.7131	2.1069	2.5361	2.6979
0.40	1.6941	2.0761	2.4901	2.6440
0.41	1.6761	2.0469	2.4464	2.5927
0.42	1.6590	2.0191	2.4047	2.5438
0.43	1.6426	1.9925	2.3650	2.4972
0.44	1.6270	1.9672	2.3271	2.4527

25. a)

a	x			
	0.3	0.5	0.8	1.0
1.0	0.76448	0.65106	0.51458	0.43876
1.5	0.75634	0.63515	0.48978	0.41079
2.0	0.75237	0.62762	0.47865	0.39867
2.5	0.75002	0.62322	0.47233	0.39189
3.0	0.74846	0.62034	0.46825	0.38756
3.5	0.74735	0.61831	0.46540	0.38456
4.0	0.74653	0.61680	0.46330	0.38235
4.5	0.74589	0.61564	0.46168	0.38066
5.0	0.74538	0.61471	0.46040	0.37932
5.5	0.74496	0.61395	0.45936	0.37823
6.0	0.74461	0.61332	0.45850	0.37734
6.5	0.74432	0.61279	0.45777	0.37658

Continuation of 25. a)

a	x			
	0.3	0.5	0.8	1.0
7.0	0.74406	0.61234	0.45715	0.37594
7.5	0.74385	0.61194	0.45662	0.37539
8.0	0.74366	0.61160	0.45615	0.37490
8.5	0.74349	0.61130	0.45574	0.37448
9.0	0.74334	0.61103	0.45538	0.37410
9.5	0.74321	0.61079	0.45505	0.37377
10.0	0.74309	0.61058	0.45476	0.37347
10.5	0.74298	0.61038	0.45450	0.37320
11.0	0.74288	0.61021	0.45426	0.37295
11.5	0.74279	0.61005	0.45404	0.37272
12.0	0.74271	0.60990	0.45384	0.37252
12.5	0.74263	0.60976	0.45366	0.37233
13.0	0.74256	0.60964	0.45349	0.37215

25. b)

a	x			
	0.3	0.5	0.8	1.0
1.0	0.74082	0.60653	0.44933	0.36788
1.5	0.66345	0.49869	0.31901	0.23391
2.0	0.63052	0.45798	0.27861	0.19788
2.5	0.61229	0.43659	0.25893	0.18114
3.0	0.60071	0.42341	0.24729	0.17146
3.5	0.59270	0.41447	0.23959	0.16517
4.0	0.58684	0.40801	0.23413	0.16074
4.5	0.58235	0.40313	0.23005	0.15745
5.0	0.57882	0.39930	0.22688	0.15492
5.5	0.57595	0.39622	0.22436	0.15291
6.0	0.57359	0.39370	0.22230	0.15127
6.5	0.57160	0.39158	0.22058	0.14991
7.0	0.56991	0.38979	0.21913	0.14877
7.5	0.56846	0.38824	0.21789	0.14779
8.0	0.56719	0.38691	0.21682	0.14695
8.5	0.56607	0.38573	0.21588	0.14621
9.0	0.56508	0.38469	0.21505	0.14556
9.5	0.56420	0.38377	0.21432	0.14498
10.0	0.56341	0.38294	0.21366	0.14447
10.5	0.56270	0.38220	0.21307	0.14400
11.0	0.56205	0.38152	0.21253	0.14359
11.5	0.56146	0.38091	0.21205	0.14321
12.0	0.56092	0.38035	0.21160	0.14286
12.5	0.56043	0.37983	0.21120	0.14254
13.0	0.55997	0.37936	0.21082	0.14225

25. c)

a	x = 1.3	1.5	1.8	2.0
−1.1	−0.29568	−0.47177	−0.72871	−0.90770
−1.0	−0.26761	−0.42422	−0.64497	−0.78949
−0.9	−0.23991	−0.37822	−0.56824	−0.68794
−0.8	−0.21252	−0.33350	−0.49646	−0.59648
−0.7	−0.18540	−0.28982	−0.42833	−0.51180
−0.6	−0.15851	−0.24699	−0.36296	−0.43191
−0.5	−0.13181	−0.20485	−0.29969	−0.35553
−0.4	−0.10527	−0.16326	−0.23802	−0.28172
−0.3	−0.07884	−0.12209	−0.17754	−0.20979
−0.2	−0.05251	−0.08123	−0.11791	−0.13917
−0.1	−0.02624	−0.04056	−0.05882	−0.06938

26. In the answers to problem 26 the first row corresponds to $y_1(x)$, the second to $y_2(x)$, etc.

	x = 0.4	0.6	0.8	1.0
a)	−0.07841	−0.17202	−0.29507	−0.44005
	0.96040	0.91200	0.84629	0.76520
b)	1.0064	1.0326	1.1042	1.2606
	1.1667	1.3965	1.7662	2.3460
c)	0.50818	0.17788	−0.22554	−0.71828
	3.37057	4.19520	5.12173	6.15485
	4.86239	6.01732	7.34727	8.87313
d)	−0.83246	−0.91974	−1.12420	−1.45404
	0.50818	0.17788	−0.22554	−0.71828
	−0.65936	−0.90238	−1.10134	−1.26424

26. e)

a	x = 0.6	1.0	a	x = 0.6	1.0
0.0	0.26069	−0.30117	0.4	0.33141	−0.44929
	1.3900	1.3818		1.7670	2.0614
0.1	0.27681	−0.33284	0.5	0.35190	−0.49654
	1.4759	1.5271		1.8763	2.2782
0.2	0.29393	−0.36785	0.6	0.37366	−0.54877
	1.5672	1.6877		1.9923	2.5178
0.3	0.31211	−0.40654	0.7	0.39676	−0.60648
	1.6641	1.8652		2.1155	2.7825

Continuation of 26. e)

a	x 0.6	x 1.0	a	x 0.6	x 1.0
0.8	0.42130 2.2463	−0.67026 3.0752	1.6	0.68085 3.6302	− 1.4917 6.8440
0.9	0.44735 2.3852	−0.74076 3.3986	1.7	0.72295 3.8547	− 1.6486 7.5638
1.0	0.47501 2.5327	−0.81866 3.7560	1.8	0.76766 4.0930	− 1.8220 8.3593
1.1	0.50439 2.6893	−0.90476 4.1511	1.9	0.81513 4.3461	− 2.0136 9.2384
1.2	0.53558 2.8556	−0.99992 4.5876	2.0	0.86554 4.6149	− 2.2253 10.210
1.3	0.56869 3.0322	−1.1051 5.0701	2.1	0.91906 4.9003	− 2.4594 11.284
1.4	0.60386 3.2197	−1.2213 5.6034	2.2	0.97590 5.2033	− 2.7180 12.471
1.5	0.64120 3.4188	−1.3497 6.1927	2.3	1.0366 5.5250	− 3.0039 13.782

26. f)

a	x 0.6	x 1.0	a	x 0.6	x 1.0
25	0.6277 0.0103	0.8891 0.5677	33	0.6243 0.0433	0.8722 0.6261
26	0.6273 0.0144	0.8869 0.5751	34	0.6239 0.0474	0.8701 0.6333
27	0.6268 0.0186	0.8848 0.5825	35	0.6235 0.0515	0.8681 0.6405
28	0.6264 0.0227	0.8827 0.5898	36	0.6231 0.0556	0.8660 0.6476
29	0.6260 0.0268	0.8805 0.5971	37	0.6226 0.0597	0.8640 0.6547
30	0.6256 0.0310	0.8784 0.6044	38	0.6222 0.0638	0.8620 0.6617
31	0.6252 0.0351	0.8763 0.6117	39	0.6218 0.0679	0.8599 0.6688
32	0.6247 0.0392	0.8743 0.6189	40	0.6214 0.0719	0.8579 0.6758

Continuation of 26. f)

a	x 0.6	x 1.0	a	x 0.6	x 1.0
41	0.6210 0.0760	0.8560 0.6828	45	0.6193 0.0923	0.8481 0.7105
42	0.6206 0.0801	0.8540 0.6897	46	0.6189 0.0963	0.8461 0.7173
43	0.6202 0.0841	0.8520 0.6967	47	0.6185 0.1003	0.8442 0.7241
44	0.6197 0.0882	0.8500 0.7036	48	0.6181 0.1044	0.8423 0.7300

26. g)

a	k	x 0.5	x 1.0	a	k	x 0.5	x 1.0
2.00	0.25	0.58954 0.55978	0.57573 0.86672	2.50	0.25	0.52698 0.52552	0.50507 0.80035
	0.50	0.57307 0.56056	0.54210 0.87614		0.50	0.51345 0.52653	0.47903 0.80958
	0.75	0.55703 0.56132	0.51032 0.88526		0.75	0.50024 0.52752	0.45418 0.81860
	1.00	0.54142 0.56206	0.48035 0.89406		1.00	0.48733 0.52849	0.43053 0.82738
	1.25	0.52622 0.56279	0.45216 0.90253		1.25	0.47473 0.52944	0.40807 0.83592
2.25	0.25	0.55640 0.54165	0.53768 0.83109	2.75	0.25	0.50067 0.51105	0.47679 0.77349
	0.50	0.54153 0.54256	0.50826 0.84046		0.50	0.48830 0.51214	0.45352 0.78254
	0.75	0.52702 0.54346	0.48030 0.84957		0.75	0.47620 0.51320	0.43124 0.79141
	1.00	0.51287 0.54433	0.45380 0.85841		1.00	0.46436 0.51424	0.40996 0.80008
	1.25	0.49907 0.54518	0.42875 0.86696		1.25	0.45279 0.51527	0.38967 0.80855

Continuation of 26. g)

a	k	x 0.5	x 1.0	a	k	x 0.5	x 1.0
3.00	0.25	0.47700	0.45201		1.00	0.44359	0.39162
		0.49801	0.74977			0.50136	0.77584
	0.50	0.46563	0.43105		1.25	0.43291	0.37317
		0.49914	0.75862			0.50244	0.78419
	0.75	0.45449	0.41092				
		0.50026	0.76731				

26. h)

a	k	x 0.2	x 0.5	x 0.8	x 1.0
1.00	1.00	1.0000	1.0000	1.0000	1.0000
		1.0000	1.0000	1.0000	1.0000
		1.0000	1.0000	1.0000	1.0000
	1.25	0.9450	0.8565	0.7854	0.7591
		1.0099	1.0580	1.1318	1.1841
		1.0426	1.0710	1.0498	1.0108
	1.50	0.8880	0.7062	0.5679	0.5264
		1.0197	1.1143	1.2537	1.3462
		1.0824	1.1223	1.0515	0.9543
	1.75	0.8292	0.5518	0.3572	0.3185
		1.0295	1.1685	1.3636	1.4824
		1.1192	1.1533	1.0065	0.8372
	2.00	0.7687	0.3957	0.1625	0.1495
		1.0392	1.2203	1.4597	1.5900
		1.1529	1.1638	0.9181	0.6706
1.25	1.00	1.0099	1.0580	1.1318	1.1841
		1.0426	1.0710	1.0498	1.0108
		0.9450	0.8565	0.7854	0.7591
	1.25	0.9563	0.9241	0.9410	0.9772
		1.0536	1.1333	1.1868	1.1978
		0.9877	0.9284	0.8403	0.7812
	1.50	0.9007	0.7831	0.7452	0.7738
		1.0644	1.1932	1.3118	1.3599
		1.0275	0.9814	0.8508	0.7430
	1.75	0.8433	0.6375	0.5537	0.5890
		1.0751	1.2503	1.4227	1.4933
		1.0644	1.0150	0.8181	0.6504

Continuation of 26. h)

a	k	x 0.2	0.5	0.8	1.0
	2.00	0.7842	0.4899	0.3753	0.4363
		1.0858	1.3044	1.5178	1.5957
		1.0983	1.0289	0.7452	0.5132
1.50	1.00	1.0197	1.1143	1.2537	1.3462
		1.0824	1.1223	1.0515	0.9543
		0.8880	0.7062	0.5679	0.5264
	1.25	0.9676	0.9904	1.0877	1.1748
		1.0944	1.1887	1.1933	1.1440
		0.9307	0.7786	0.6268	0.5579
	1.50	0.9134	0.8591	0.9149	1.0021
		1.1062	1.2522	1.3216	1.3069
		0.9705	0.8330	0.6453	0.5366
	1.75	0.8575	0.7230	0.7439	0.8423
		1.1179	1.3123	1.4341	1.4394
		1.0075	0.8690	0.6244	0.4676
	2.00	0.7998	0.5843	0.5831	0.7077
		1.1294	1.3687	1.5293	1.5394
		1.0416	0.8862	0.5668	0.3596
1.75	1.00	1.0295	1.1685	1.3636	1.4824
		1.1192	1.1533	1.0065	0.8372
		0.8292	0.5518	0.3572	0.3185
	1.25	0.9788	1.0550	1.2230	1.3466
		1.1322	1.2237	1.1528	1.0292
		0.8718	0.6242	0.4188	0.3570
	1.50	0.9262	0.9338	1.0739	1.2052
		1.1450	1.2907	1.2843	1.1936
		0.9117	0.6797	0.4443	0.3505
	1.75	0.8717	0.8075	0.9242	1.0712
		1.1576	1.3538	1.3990	1.3268
		0.9487	0.7178	0.4344	0.3035
	2.00	0.8154	0.6782	0.7818	0.9558
		1.1701	1.4126	1.4952	1.4271
		0.9829	0.7380	0.3915	0.2231

Continuation of 26. h)

a	k	x			
		0.2	0.5	0.8	1.0
2.00	1.00	1.0392	1.2203	1.4597	1.5900
		1.1529	1.1638	0.9181	0.6706
		0.7687	0.3957	0.1625	0.1495
	1.25	0.9901	1.1173	1.3446	1.4887
		1.1669	1.2380	1.0683	0.8641
		0.8112	0.4678	0.2255	0.1924
	1.50	0.9389	1.0066	1.2193	1.3782
		1.1807	1.3085	1.2030	1.0301
		0.8511	0.5241	0.2568	0.1984
	1.75	0.8859	0.8903	1.0914	1.2699
		1.1943	1.3745	1.3202	1.1654
		0.8882	0.5639	0.2569	0.1707
	2.00	0.8312	0.7708	0.9679	1.1741
		1.2076	1.4357	1.4182	1.2682
		0.9224	0.5870	0.2278	0.1156

26. i)

a	k	x			
		0.2	0.5	0.8	1.0
1.0	1.5	0.97802	0.89381	0.75871	0.64905
		−0.07380	−0.11807	−0.07816	−0.00698
		0.40932	0.87895	1.29967	1.52473
	1.7	0.97387	0.87258	0.71088	0.58103
		−0.06325	−0.09226	−0.03934	0.03809
		0.44824	0.96519	1.40806	1.62912
	1.9	0.96932	0.84918	0.65858	0.50741
		−0.05490	−0.07186	−0.00884	0.07303
		0.48698	1.04916	1.50822	1.71885
	2.1	0.96439	0.82366	0.60214	0.42896
		−0.04814	−0.05532	0.01564	0.10055
		0.52553	1.13064	1.59946	1.79291
	2.3	0.95907	0.79609	0.54193	0.34648
		−0.04256	−0.04165	0.03560	0.12245
		0.56387	1.20944	1.68116	1.85042

Continuation of 26. i)

a	k	x = 0.2	0.5	0.8	1.0
1.1	1.5	0.97871	0.89581	0.76327	0.65674
		−0.07949	−0.11892	−0.05850	0.03344
		0.44149	0.96102	1.42129	1.66105
	1.7	0.97465	0.87486	0.71614	0.58996
		−0.06789	−0.09075	−0.01674	0.08112
		0.48428	1.05538	1.53841	1.77203
	1.9	0.97020	0.85174	0.66456	0.51763
		−0.05872	−0.06849	0.01599	0.11791
		0.52689	1.14727	1.64659	1.86715
	2.1	0.96535	0.82650	0.60886	0.44050
		−0.05129	−0.05046	0.04219	0.14671
		0.56929	1.23646	1.74508	1.94532
	2.3	0.96013	0.79921	0.54940	0.35938
		−0.04515	−0.03557	0.06347	0.16943
		0.61146	1.32271	1.83321	2.00560
1.2	1.5	0.97929	0.89757	0.76757	0.66429
		−0.08485	−0.11769	−0.03388	0.08081
		0.47383	1.04310	1.54045	1.79127
	1.7	0.97530	0.87686	0.72112	0.59876
		−0.07221	−0.08722	0.01057	0.13066
		0.52048	1.14539	1.66561	1.90767
	1.9	0.97093	0.85399	0.67025	0.52775
		−0.06222	−0.06316	0.04533	0.16892
		0.56694	1.24501	1.78115	2.00708
	2.1	0.96616	0.82901	0.61529	0.45200
		−0.05413	−0.04367	0.07307	0.19867
		0.61317	1.34171	1.88626	2.08834
	2.3	0.96101	0.80198	0.55659	0.37228
		−0.04744	−0.02760	0.09553	0.22191
		0.65917	1.43525	1.98020	2.15043
1.3	1.5	0.97978	0.89913	0.77170	0.67178
		−0.08988	−0.11439	−0.00442	0.13481
		0.50632	1.12504	1.65658	1.91436
	1.7	0.97586	0.87865	0.72593	0.60754
		−0.07620	−0.08168	0.04247	0.18636
		0.55683	1.23505	1.78905	2.03496

Continuation of 26. i)

a	k	x = 0.2	0.5	0.8	1.0
	1.9	0.97155	0.85602	0.67578	0.53789
		−0.06540	−0.05586	0.07906	0.22569
		0.60712	1.34220	1.91123	2.13751
	2.1	0.96685	0.83127	0.62156	0.46356
		−0.05665	−0.03497	0.10816	0.25604
		0.65717	1.44622	2.02227	2.22078
	2.3	0.96177	0.80449	0.56364	0.38533
		−0.04941	−0.01775	0.13163	0.27951
		0.70697	1.54683	2.12137	2.28369
1.4	1.5	0.98020	0.90056	0.77572	0.67929
		−0.09459	−0.10902	0.02975	0.19501
		0.53896	1.20669	1.76908	2.02936
	1.7	0.97634	0.88029	0.73064	0.61637
		−0.07988	−0.07414	0.07881	0.24778
		0.59330	1.32420	1.90809	2.15289
	1.9	0.97209	0.85787	0.68121	0.54813
		−0.06826	−0.04662	0.11700	0.28780
		0.64741	1.43865	2.03618	2.25739
	2.1	0.96745	0.83336	0.62775	0.47528
		−0.05885	−0.02437	0.14728	0.31841
		0.70127	1.54976	2.15243	2.34155
	2.3	0.96242	0.80680	0.57062	0.39859
		−0.05107	−0.00604	0.17160	0.34180
		0.75486	1.65724	2.25601	2.40425

26. j)

a	k	x = 0.2	0.5	0.8	1.0
1.00	2.00	1.2013	1.4041	1.3306	1.1540
		1.2902	1.5675	1.4682	1.2237
		0.7463	0.1689	−0.4792	−0.8175
	2.25	1.2410	1.4137	1.2092	0.9643
		1.2744	1.4676	1.2385	0.9586
		0.6782	−0.0372	−0.7334	−1.0345
	2.50	1.2766	1.3949	1.0610	0.7665
		1.2600	1.3722	1.0570	0.7842
		0.6086	−0.2326	−0.9351	−1.1885

Continuation of 26. j)

a	k	x: 0.2	0.5	0.8	1.0
	2.75	1.3081	1.3521	0.8974	0.5620
		1.2466	1.2827	0.9210	0.6833
		0.5375	−0.4146	−1.0930	−1.2977
	3.00	1.3355	1.2891	0.7234	0.3456
		1.2338	1.2005	0.8260	0.6426
		0.4653	−0.5815	−1.2152	−1.3713
1.25	2.00	1.1292	1.3190	1.4133	1.3877
		1.3211	1.7331	1.9239	1.8728
		0.8514	0.5101	0.0501	−0.2724
	2.25	1.1666	1.3730	1.3948	1.2894
		1.3045	1.6496	1.6848	1.5127
		0.7994	0.3388	−0.2333	−0.5809
	2.50	1.2013	1.4041	1.3306	1.1540
		1.2902	1.5675	1.4682	1.2237
		0.7463	0.1689	−0.4792	−0.8175
	2.75	1.2334	1.4142	1.2361	1.0031
		1.2774	1.4873	1.2806	1.0038
		0.6920	0.0033	−0.6870	−0.9969
	3.00	1.2628	1.4056	1.1226	0.8463
		1.2656	1.4097	1.1239	0.8443
		0.6366	−0.1560	−0.8601	−1.1331
1.50	2.00	1.0752	1.2079	1.3329	1.3903
		1.3487	1.8467	2.2533	2.4286
		0.9187	0.7354	0.4726	0.2590
	2.25	1.1095	1.2828	1.3994	1.4115
		1.3305	1.7753	2.0480	2.0754
		0.8769	0.5954	0.2041	−0.0883
	2.50	1.1420	1.3397	1.4134	1.3611
		1.3153	1.7051	1.8424	1.7453
		0.8342	0.4530	−0.0482	−0.3839
	2.75	1.1726	1.3798	1.3868	1.2686
		1.3020	1.6359	1.6469	1.4595
		0.7906	0.3103	−0.2770	−0.6250
	3.00	1.2013	1.4041	1.3306	1.1540
		1.2902	1.5675	1.4682	1.2237
		0.7463	0.1689	−0.4792	−0.8175

Continuation of 26. j)

a	k	x 0.2	0.5	0.8	1.0
1.75	2.00	1.0333	1.0991	1.1781	1.2337
		1.3747	1.9317	2.4682	2.8019
		0.9655	0.8899	0.7823	0.6914
	2.25	1.0650	1.1831	1.3026	1.3662
		1.3546	1.8675	2.3099	2.5281
		0.9305	0.7747	0.5507	0.3654
	2.50	1.0951	1.2530	1.3780	1.4131
		1.3379	1.8057	2.1368	2.2259
		0.8949	0.6558	0.3178	0.0549
	2.75	1.1237	1.3093	1.4111	1.3967
		1.3237	1.7451	1.9592	1.9294
		0.8587	0.5346	0.0934	−0.2218
	3.00	1.1509	1.3528	1.4094	1.3379
		1.3113	1.6852	1.7851	1.6587
		0.8218	0.4122	−0.1162	−0.4582
2.00	2.00	1.0000	1.0000	1.0000	1.0000
		1.4000	2.0000	2.6000	3.0000
		1.0000	1.0000	1.0000	1.0000
	2.25	1.0293	1.0876	1.1589	1.2102
		1.3776	1.9399	2.4864	2.8318
		0.9698	0.9040	0.8106	0.7318
	2.50	1.0572	1.1635	1.2763	1.3419
		1.3592	1.8833	2.3513	2.6009
		0.9393	0.8039	0.6091	0.4465
	2.75	1.0839	1.2284	1.3551	1.4042
		1.3438	1.8287	2.2027	2.3401
		0.9083	0.7008	0.4046	0.1680
	3.00	1.1095	1.2828	1.3994	1.4115
		1.3305	1.7753	2.0480	2.0754
		0.8769	0.5954	0.2041	−0.0883

26. k)

a	k	x = 0.2	x = 0.5	x = 0.8	x = 1.0
1.00	1.25	1.2157	1.6068	2.1089	2.5357
		1.2832	1.8506	2.6228	3.2687
		1.2280	1.7077	2.4328	3.1083
	1.50	1.2097	1.5610	1.9736	2.3139
		1.3482	2.0794	3.1081	3.9759
		1.2347	1.7717	2.6685	3.5655
	1.75	1.2036	1.5109	1.8163	2.0436
		1.4164	2.3388	3.7016	4.8861
		1.2417	1.8412	2.9370	4.1037
	2.00	1.1972	1.4561	1.6330	1.7135
		1.4882	2.6327	4.4277	6.0594
		1.2490	1.9166	3.2439	4.7398
	2.25	1.1906	1.3961	1.4192	1.3097
		1.5637	2.9659	5.3165	7.5735
		1.2565	1.9987	3.5952	5.4943
1.30	1.25	1.2981	1.9248	2.8563	3.7131
		1.2842	1.8720	2.7398	3.5416
		1.2956	1.9025	2.7804	3.5735
	1.50	1.2921	1.8769	2.7124	3.4761
		1.3492	2.1008	3.2250	4.2485
		1.3025	1.9699	3.0373	4.0837
	1.75	1.2858	1.8246	2.5455	3.1881
		1.4174	2.3601	3.8177	5.1562
		1.3096	2.0431	3.3292	4.6814
	2.00	1.2793	1.7675	2.3514	2.8375
		1.4892	2.6539	4.5424	6.3243
		1.3170	2.1225	3.6618	5.3844
	2.25	1.2726	1.7049	2.1255	2.4094
		1.5647	2.9870	5.4289	7.8296
		1.3247	2.2087	4.0415	6.2145
1.60	1.25	1.3858	2.2969	3.8180	5.3233
		1.2853	1.8969	2.8880	3.9077
		1.3672	2.1242	3.1929	4.1255
	1.50	1.3797	2.2468	3.6646	5.0691
		1.3502	2.1257	3.3737	4.6167
		1.3742	2.1954	3.4740	4.6983

Continuation of 26. k)

a	k	x 0.2	0.5	0.8	1.0
	1.75	1.3733	2.1922	3.4869	4.7613
		1.4185	2.3850	3.9663	5.5245
		1.3815	2.2726	3.7925	5.3660
	2.00	1.3667	2.1325	3.2810	4.3875
		1.4903	2.6787	4.6902	6.6899
		1.3891	2.3562	4.1542	6.1473
	2.25	1.3598	2.0673	3.0416	3.9325
		1.5658	3.0117	5.5751	8.1889
		1.3969	2.4469	4.5660	7.0657
1.90	1.25	1.4791	2.7321	5.0535	7.5206
		1.2864	1.9257	3.0749	4.3954
		1.4430	2.3770	3.6839	4.7806
	1.50	1.4728	2.6796	4.8894	7.2469
		1.3513	2.1546	3.5619	5.1099
		1.4502	2.4522	3.9926	5.4277
	1.75	1.4663	2.6225	4.6999	6.9165
		1.4196	2.4139	4.1553	6.0212
		1.4577	2.5337	4.3413	6.1780
	2.00	1.4596	2.5601	4.4806	6.5167
		1.4914	2.7076	4.8792	7.1873
		1.4654	2.6218	4.7361	7.0516
	2.25	1.4526	2.4920	4.2265	6.0314
		1.5669	3.0405	5.7633	8.6834
		1.4733	2.7174	5.1842	8.0734
2.20	1.25	1.5783	3.2408	6.6385	10.514
		1.2876	1.9590	3.3098	5.0419
		1.5234	2.6654	4.2696	5.5576
	1.50	1.5719	3.1858	6.4625	10.218
		1.3525	2.1880	3.7991	5.7665
		1.5308	2.7450	4.6101	6.2932
	1.75	1.5652	3.1259	6.2598	9.8618
		1.4208	2.4474	4.3943	6.6857
		1.5384	2.8311	4.9933	7.1416

Continuation of 26. k)

a	k	x = 0.2	0.5	0.8	1.0
	2.00	1.5584	3.0606	6.0257	9.4324
		1.4926	2.7412	5.1193	7.8570
		1.5462	2.9242	5.4258	8.1244
	2.25	1.5513	2.9894	5.7551	8.9130
		1.5681	3.0741	6.0037	9.3550
		1.5544	3.0250	5.9153	9.2680

27. a)

a	x	k = 0.3	0.4	0.5	0.6	0.7
−0.3	0.5	−0.02709	−0.02761	−0.02813	−0.02866	−0.02918
	0.8	0.06839	0.06497	0.06155	0.05814	0.05472
	1.0	0.17204	0.16380	0.15555	0.14730	0.13904
−0.1	0.5	0.07334	0.07282	0.07230	0.07178	0.07126
	0.8	0.22652	0.22316	0.21981	0.21645	0.21310
	1.0	0.36089	0.35290	0.34491	0.33692	0.32892
0.1	0.5	0.17212	0.17160	0.17108	0.17057	0.17005
	0.8	0.37492	0.37161	0.36830	0.36499	0.36169
	1.0	0.52999	0.52213	0.51426	0.50640	0.49854
0.3	0.5	0.26937	0.26885	0.26834	0.26782	0.26731
	1.0	0.68762	0.67974	0.67187	0.66400	0.65614
0.5	1.0	0.84188	0.83388	0.82589	0.81790	0.80992

27. b)

a	x	k = 1.6	1.8	2.0	2.2	2.4
−0.3	0.5	−0.05594	−0.05585	−0.05576	−0.05566	−0.05555
	0.8	−0.04705	−0.04680	−0.04652	−0.04622	−0.04589
	1.0	−0.04856	−0.04818	−0.04775	−0.04729	−0.04679
−0.1	0.5	0.04377	0.04377	0.04378	0.04378	0.04379
	0.8	0.11228	0.11235	0.11243	0.11252	0.11262
	1.0	0.15072	0.15092	0.15113	0.15137	0.15162
0.1	0.5	0.14421	0.14433	0.14447	0.14462	0.14478
	0.8	0.27519	0.27603	0.27695	0.27796	0.27906
	1.0	0.35669	0.35843	0.36034	0.36243	0.36468
0.3	0.5	0.24539	0.24581	0.24629	0.24681	0.24737
	1.0	0.56869	0.57329	0.57825	0.58350	0.58898
0.5	1.0	0.78530	0.79327	0.80152	0.80983	0.81804

27. c)

a	x	k = 0.7	0.8	0.9	1.0	1.1
−0.3	0.5	−0.02869	−0.02921	−0.02973	−0.03025	−0.03077
	0.8	0.05595	0.05254	0.04912	0.04571	0.04230
	1.0	0.14084	0.13257	0.12430	0.11603	0.10775
−0.1	0.5	0.07134	0.07082	0.07030	0.06978	0.06925
	0.8	0.21481	0.21143	0.20804	0.20465	0.20127
	1.0	0.33323	0.32515	0.31707	0.30898	0.30089
0.1	0.5	0.17117	0.17065	0.17013	0.16961	0.16909
	0.8	0.37066	0.36730	0.36394	0.36058	0.35722
	1.0	0.51551	0.50757	0.49963	0.49168	0.48374
0.3	0.5	0.27082	0.27030	0.26978	0.26926	0.26874
	1.0	0.69005	0.68219	0.67433	0.66647	0.65861
0.5	1.0	0.85955	0.85171	0.84387	0.83603	0.82820

27. d)

a	x	k = 0.8	0.9	1.0	1.1	1.2
−0.3	0.5	−0.02314	−0.02291	−0.02268	−0.02245	−0.02223
	0.8	0.08463	0.08517	0.08570	0.08623	0.08674
	1.0	0.19929	0.19909	0.19887	0.19863	0.19836
−0.1	0.5	0.07438	0.07430	0.07422	0.07414	0.07407
	0.8	0.22883	0.22746	0.22609	0.22473	0.22338
	1.0	0.36284	0.35835	0.35389	0.34947	0.34509
0.1	0.5	0.17190	0.17152	0.17113	0.17075	0.17036
	0.8	0.37349	0.37025	0.36704	0.36384	0.36066
	1.0	0.52950	0.52100	0.51256	0.50419	0.49590
0.3	0.5	0.26944	0.26875	0.26806	0.26738	0.26669
	1.0	0.70103	0.68895	0.67694	0.66501	0.65316
0.5	1.0	0.87870	0.86365	0.84864	0.83367	0.81875

27. e)

a	x	k = 0.5	0.6	0.7	0.8	0.9
−0.3	0.5	−0.04277	−0.04297	−0.04316	−0.04336	−0.04356
	0.8	0.01531	0.01476	0.01421	0.01365	0.01310
	1.0	0.08254	0.08177	0.08100	0.08021	0.07942
−0.1	0.5	0.05827	0.05829	0.05830	0.05832	0.05833
	0.8	0.17961	0.17993	0.18024	0.18056	0.18088
	1.0	0.29098	0.29192	0.29286	0.29380	0.29474

Continuation of 27. e)

a	x	k				
		0.5	0.6	0.7	0.8	0.9
0.1	0.5	0.15932	0.15954	0.15976	0.15999	0.16021
	0.8	0.34390	0.34508	0.34627	0.34745	0.34864
	1.0	0.49939	0.50202	0.50465	0.50729	0.50992
0.3	0.5	0.26036	0.26079	0.26123	0.26166	0.26209
	1.0	0.70775	0.71204	0.71631	0.72056	0.72479
0.5	1.0	0.91603	0.92192	0.92776	0.93352	0.93921

27. f)

a	x	k				
		0.5	0.6	0.7	0.8	0.9
−0.3	0.5	−0.02905	−0.02905	−0.02906	−0.02907	−0.02909
	0.8	0.05661	0.05659	0.05657	0.05654	0.05651
	1.0	0.14695	0.14692	0.14687	0.14682	0.14677
−0.1	0.5	0.07097	0.07096	0.07096	0.07096	0.07096
	0.8	0.21655	0.21651	0.21646	0.21640	0.21634
	1.0	0.34661	0.34642	0.34620	0.34594	0.34565
0.1	0.5	0.17093	0.17091	0.17089	0.17086	0.17083
	0.8	0.37616	0.37595	0.37569	0.37540	0.37507
	1.0	0.54543	0.54473	0.54390	0.54295	0.54188
0.3	0.5	0.27084	0.27078	0.27071	0.27063	0.27054
	1.0	0.74344	0.74187	0.74003	0.73792	0.73556
0.5	1.0	0.94064	0.93786	0.93462	0.93092	0.92678

27. g)

a	x			
	0.3	0.5	0.8	1.0
0.2	0.19993	0.19980	0.19949	0.19920
0.3	0.19989	0.19970	0.19923	0.19880
0.4	0.19986	0.19960	0.19898	0.19841
0.5	0.19982	0.19950	0.19872	0.19801
0.6	0.19978	0.19940	0.19847	0.19761
0.7	0.19975	0.19930	0.19822	0.19722
0.8	0.19971	0.19920	0.19796	0.19683
0.9	0.19968	0.19910	0.19771	0.19643
1.0	0.19964	0.19900	0.19746	0.19604
1.1	0.19960	0.19890	0.19720	0.19565
1.2	0.19957	0.19880	0.19695	0.19526
1.3	0.19953	0.19870	0.19670	0.19487
1.4	0.19950	0.19860	0.19645	0.19448

Continuation of 27. g)

a	x 0.3	0.5	0.8	1.0
1.5	0.19946	0.19851	0.19620	0.19409
1.6	0.19942	0.19841	0.19595	0.19370
1.7	0.19939	0.19831	0.19569	0.19331
1.8	0.19935	0.19821	0.19544	0.19293
1.9	0.19932	0.19811	0.19519	0.19254
2.0	0.19928	0.19801	0.19494	0.19216
2.1	0.19925	0.19791	0.19470	0.19177
2.2	0.19921	0.19781	0.19445	0.19139
2.3	0.19917	0.19771	0.19420	0.19101
2.4	0.19914	0.19761	0.19395	0.19062
2.5	0.19910	0.19752	0.19370	0.19024
2.6	0.19907	0.19742	0.19345	0.18986

27. h)

n	x 0.4	0.6	0.8	1.0
1.0	0.97355	0.94107	0.89670	0.84147
1.5	0.97365	0.94159	0.89828	0.84517
2.0	0.97375	0.94209	0.89980	0.84865
2.5	0.97386	0.94259	0.90126	0.85194
3.0	0.97396	0.94307	0.90267	0.85506
3.5	0.97405	0.94354	0.90403	0.85801
4.0	0.97415	0.94401	0.90535	0.86082
4.5	0.97425	0.94446	0.90661	0.86348
5.0	0.97435	0.94491	0.90784	0.86603

28. a)

a	x 0.2	0.4	0.6	0.8
0.5	0.03394	0.05921	0.06732	0.05014
1.0	0.03610	0.06278	0.07102	0.05250
1.5	0.03853	0.06679	0.07515	0.05514
2.0	0.04127	0.07132	0.07980	0.05810
2.5	0.04440	0.07646	0.08508	0.06145
3.0	0.04800	0.08237	0.09112	0.06528
3.5	0.05217	0.08921	0.09809	0.06968
4.0	0.05707	0.09723	0.10625	0.07482
4.5	0.06289	0.10675	0.11592	0.08089
5.0	0.06993	0.11825	0.12756	0.08819
5.5	0.07861	0.13239	0.14186	0.09713
6.0	0.08955	0.15020	0.15983	0.10834

28. b)

i	x	a = 0.70	0.75	0.80	0.85	0.90
1	0.2	0.20261	0.20249	0.20237	0.20224	0.20211
	0.4	0.31383	0.31358	0.31331	0.31303	0.31273
	0.6	0.31995	0.31959	0.31922	0.31882	0.31840
	0.8	0.21436	0.21406	0.21375	0.21341	0.21306
2	0.2	0.20504	0.20492	0.20480	0.20467	0.20453
	0.4	0.31583	0.31558	0.31530	0.31501	0.31471
	0.6	0.31972	0.31936	0.31899	0.31859	0.31817
	0.8	0.21231	0.21201	0.21170	0.21136	0.21101
3	0.2	0.20721	0.20709	0.20696	0.20682	0.20668
	0.4	0.31756	0.31729	0.31702	0.31672	0.31641
	0.6	0.31944	0.31908	0.31870	0.31830	0.31788
	0.8	0.21046	0.21016	0.20985	0.20952	0.20917
4	0.2	0.20915	0.20902	0.20889	0.20875	0.20861
	0.4	0.31905	0.31878	0.31850	0.31821	0.31789
	0.6	0.31913	0.31877	0.31839	0.31798	0.31756
	0.8	0.20879	0.20850	0.20818	0.20785	0.20751
5	0.2	0.21089	0.21077	0.21064	0.21049	0.21035
	0.4	0.32036	0.32009	0.31980	0.31950	0.31918
	0.6	0.31879	0.31843	0.31805	0.31764	0.31722
	0.8	0.20727	0.20698	0.20667	0.20634	0.20600

28. c)

i	x	a = 0.2	0.3	0.4	0.5	0.6
1	0.2	0.15809	0.15729	0.15650	0.15573	0.15498
	0.4	0.23649	0.23518	0.23390	0.23266	0.23144
	0.6	0.23577	0.23443	0.23313	0.23186	0.23063
	0.8	0.15675	0.15592	0.15511	0.15432	0.15355
2	0.2	0.15812	0.15731	0.15653	0.15576	0.15501
	0.4	0.23655	0.23525	0.23397	0.23272	0.23151
	0.6	0.23588	0.23455	0.23324	0.23197	0.23074
	0.8	0.15687	0.15603	0.15522	0.15443	0.15366
3	0.2	0.15814	0.15734	0.15655	0.15578	0.15503
	0.4	0.23662	0.23531	0.23403	0.23278	0.23157
	0.6	0.23599	0.23466	0.23335	0.23208	0.23084
	0.8	0.15698	0.15615	0.15533	0.15454	0.15377

Continuation of 28. c)

i	x	a: 0.2	0.3	0.4	0.5	0.6
4	0.2	0.15816	0.15736	0.15657	0.15580	0.15505
	0.4	0.23668	0.23538	0.23409	0.23285	0.23163
	0.6	0.23610	0.23477	0.23346	0.23218	0.23095
	0.8	0.15709	0.15626	0.15544	0.15465	0.15388
5	0.2	0.15818	0.15738	0.15659	0.15582	0.15507
	0.4	0.23675	0.23544	0.23416	0.23291	0.23169
	0.6	0.23621	0.23487	0.23356	0.23229	0.23105
	0.8	0.15721	0.15637	0.15555	0.15476	0.15399

28. d)

i	x	a: 0.5	0.6	0.7	0.8	0.9
1	0.2	0.18882	0.18844	0.18802	0.18755	0.18704
	0.4	0.28901	0.28825	0.28738	0.28642	0.28538
	0.6	0.29100	0.29002	0.28890	0.28767	0.28636
	0.8	0.19260	0.19184	0.19098	0.19004	0.18904
2	0.2	0.18883	0.18846	0.18804	0.18756	0.18705
	0.4	0.28897	0.28820	0.28733	0.28637	0.28534
	0.6	0.29082	0.28984	0.28872	0.28750	0.28618
	0.8	0.19236	0.19160	0.19074	0.18980	0.18880
3	0.2	0.18885	0.18848	0.18805	0.18758	0.18707
	0.4	0.28892	0.28816	0.28729	0.28633	0.28529
	0.6	0.29065	0.28967	0.28855	0.28732	0.28601
	0.8	0.19212	0.19136	0.19050	0.18956	0.18857
4	0.2	0.18887	0.18850	0.18807	0.18760	0.18709
	0.4	0.28888	0.28811	0.28725	0.28629	0.28525
	0.6	0.29048	0.28950	0.28838	0.28716	0.28584
	0.8	0.19189	0.19113	0.19027	0.18934	0.18834
5	0.2	0.18889	0.18852	0.18809	0.18762	0.18711
	0.4	0.28883	0.28807	0.28720	0.28624	0.28521
	0.6	0.29031	0.28933	0.28822	0.28699	0.28568
	0.8	0.19166	0.19090	0.19005	0.18911	0.18812

28. e)

i	x	a: 1.0	1.1	1.2	1.3	1.4
1	0.2	0.15691	0.15663	0.15637	0.15614	0.15592
	0.4	0.23327	0.23281	0.23239	0.23200	0.23165
	0.6	0.23198	0.23152	0.23109	0.23071	0.23037
	0.8	0.15447	0.15418	0.15393	0.15369	0.15349
2	0.2	0.15678	0.15650	0.15624	0.15601	0.15579
	0.4	0.23316	0.23270	0.23228	0.23190	0.23154
	0.6	0.23194	0.23147	0.23105	0.23067	0.23032
	0.8	0.15447	0.15418	0.15392	0.15369	0.15348
3	0.2	0.15667	0.15639	0.15613	0.15590	0.15568
	0.4	0.23307	0.23261	0.23219	0.23180	0.23145
	0.6	0.23190	0.23143	0.23101	0.23063	0.23028
	0.8	0.15446	0.15417	0.15391	0.15368	0.15348
4	0.2	0.15657	0.15630	0.15604	0.15581	0.15559
	0.4	0.23299	0.23253	0.23211	0.23172	0.23137
	0.6	0.23186	0.23139	0.23097	0.23059	0.23025
	0.8	0.15446	0.15417	0.15391	0.15368	0.15347
5	0.2	0.15649	0.15621	0.15596	0.15572	0.15551
	0.4	0.23292	0.23246	0.23204	0.23166	0.23131
	0.6	0.23183	0.23136	0.23094	0.23056	0.23022
	0.8	0.15445	0.15416	0.15390	0.15367	0.15347

28. f)

i	x	a: 0.2	0.3	0.4	0.5	0.6
1	0.2	0.14294	0.14405	0.14534	0.14683	0.14860
	0.4	0.21470	0.21638	0.21833	0.22061	0.22331
	0.6	0.21631	0.21795	0.21987	0.22213	0.22485
	0.8	0.14583	0.14681	0.14798	0.14938	0.15109
2	0.2	0.14234	0.14346	0.14474	0.14624	0.14801
	0.4	0.21489	0.21658	0.21854	0.22083	0.22356
	0.6	0.21739	0.21905	0.22099	0.22328	0.22603
	0.8	0.14705	0.14805	0.14923	0.15065	0.15239
3	0.2	0.14176	0.14287	0.14416	0.14566	0.14742
	0.4	0.21503	0.21673	0.21870	0.22101	0.22375
	0.6	0.21838	0.22006	0.22202	0.22433	0.22711
	0.8	0.14818	0.14920	0.15040	0.15185	0.15361

Continuation of 28. f)

i	x	a 0.2	0.3	0.4	0.5	0.6
4	0.2	0.14119	0.14230	0.14358	0.14508	0.14685
	0.4	0.21513	0.21684	0.21882	0.22114	0.22389
	0.6	0.21929	0.22098	0.22296	0.22530	0.22811
	0.8	0.14925	0.15028	0.15150	0.15296	0.15475
5	0.2	0.14063	0.14174	0.14302	0.14452	0.14629
	0.4	0.21519	0.21691	0.21890	0.22123	0.22399
	0.6	0.22013	0.22183	0.22383	0.22619	0.22902
	0.8	0.15025	0.15129	0.15253	0.15401	0.15582

28. g)

i	x	a 0.6	0.7	0.8	0.9	1.0
1	0.2	0.17540	0.17562	0.17588	0.17617	0.17649
	0.4	0.24654	0.24694	0.24741	0.24793	0.24851
	0.6	0.23076	0.23122	0.23175	0.23235	0.23302
	0.8	0.14373	0.14406	0.14443	0.14485	0.14531
2	0.2	0.17544	0.17567	0.17592	0.17621	0.17654
	0.4	0.24669	0.24709	0.24755	0.24807	0.24865
	0.6	0.23101	0.23148	0.23201	0.23261	0.23328
	0.8	0.14399	0.14432	0.14469	0.14511	0.14558
3	0.2	0.17548	0.17571	0.17596	0.17625	0.17658
	0.4	0.24682	0.24722	0.24769	0.24821	0.24879
	0.6	0.23126	0.23172	0.23226	0.23286	0.23353
	0.8	0.14424	0.14457	0.14494	0.14536	0.14583
4	0.2	0.17552	0.17574	0.17600	0.17629	0.17661
	0.4	0.24695	0.24735	0.24782	0.24834	0.24892
	0.6	0.23149	0.23196	0.23249	0.23310	0.23377
	0.8	0.14448	0.14481	0.14518	0.14560	0.14607
5	0.2	0.17555	0.17577	0.17603	0.17632	0.17665
	0.4	0.24707	0.24748	0.24794	0.24846	0.24905
	0.6	0.23171	0.23218	0.23272	0.23332	0.23399
	0.8	0.14471	0.14504	0.14541	0.14583	0.14630

28. h)

a	x = 0.0	0.2	0.4	0.6	0.8	1.0
8	−1.25874	−1.13927	−1.00291	−0.86417	−0.72844	−0.59498
9	−1.24994	−1.13121	−0.99563	−0.85774	−0.72294	−0.59046
10	−1.24127	−1.12326	−0.98845	−0.85140	−0.71751	−0.58601
11	−1.23273	−1.11543	−0.98138	−0.84516	−0.71217	−0.58162
12	−1.22430	−1.10771	−0.97441	−0.83900	−0.70690	−0.57730
13	−1.21600	−1.10010	−0.96754	−0.83294	−0.70170	−0.57303
14	−1.20782	−1.09260	−0.96077	−0.82696	−0.69658	−0.56883
15	−1.19975	−1.08520	−0.95409	−0.82106	−0.69154	−0.56469
16	−1.19179	−1.07791	−0.94750	−0.81525	−0.68656	−0.56060
17	−1.18394	−1.07071	−0.94101	−0.80952	−0.68165	−0.55658
18	−1.17620	−1.06362	−0.93460	−0.80386	−0.67681	−0.55261
19	−1.16857	−1.05662	−0.92829	−0.79829	−0.67204	−0.54869
20	−1.16103	−1.04972	−0.92206	−0.79279	−0.66733	−0.54482
21	−1.15361	−1.04291	−0.91591	−0.78736	−0.66269	−0.54101
22	−1.14628	−1.03620	−0.90985	−0.78201	−0.65811	−0.53725
23	−1.13904	−1.02957	−0.90386	−0.77673	−0.65359	−0.53354
24	−1.13191	−1.02303	−0.89796	−0.77152	−0.64913	−0.52988
25	−1.12486	−1.01657	−0.89213	−0.76638	−0.64473	−0.52627
26	−1.11791	−1.01020	−0.88638	−0.76130	−0.64038	−0.52271
27	−1.11105	−1.00391	−0.88070	−0.75629	−0.63610	−0.51919
28	−1.10428	−0.99771	−0.87510	−0.75135	−0.63186	−0.51571
29	−1.09759	−0.99158	−0.86957	−0.74647	−0.62769	−0.51229
30	−1.09099	−0.98553	−0.86411	−0.74165	−0.62356	−0.50890
31	−1.08447	−0.97955	−0.85871	−0.73689	−0.61949	−0.50556
32	−1.07803	−0.97365	−0.85339	−0.73219	−0.61547	−0.50226
33	−1.07168	−0.96783	−0.84813	−0.72755	−0.61150	−0.49901
34	−1.06540	−0.96208	−0.84294	−0.72297	−0.60758	−0.49579
35	−1.05920	−0.95639	−0.83781	−0.71845	−0.60371	−0.49261
36	−1.05307	−0.95078	−0.83274	−0.71398	−0.59989	−0.48947
37	−1.04702	−0.94523	−0.82774	−0.70956	−0.59611	−0.48637
38	−1.04104	−0.93976	−0.82279	−0.70520	−0.59238	−0.48331
39	−1.03514	−0.93434	−0.81791	−0.70089	−0.58869	−0.48029
40	−1.02930	−0.92900	−0.81308	−0.69664	−0.58505	−0.47730
41	−1.02354	−0.92371	−0.80832	−0.69243	−0.58145	−0.47435
42	−1.01784	−0.91849	−0.80360	−0.68828	−0.57790	−0.47143
43	−1.01221	−0.91333	−0.79895	−0.68417	−0.57438	−0.46855
44	−1.00664	−0.90823	−0.79434	−0.68011	−0.57091	−0.46570
45	−1.00114	−0.90319	−0.78979	−0.67610	−0.56748	−0.46288
46	−0.99570	−0.89820	−0.78530	−0.67213	−0.56409	−0.46010
47	−0.99032	−0.89328	−0.78085	−0.66821	−0.56074	−0.45735
48	−0.98301	−0.88841	−0.77646	−0.66434	−0.55742	−0.45463

Bei Fragen zur Produktsicherheit wenden Sie sich bitte an:
If you have any questions regarding product safety,
please contact:

Walter de Gruyter GmbH
Genthiner Straße 13
10785 Berlin
productsafety@degruyterbrill.com